# Wheat Montana

# Cookbook

Recipes from our bakery and our
customers using Wheat Montana products.

**ThreeForks™**

Published by ThreeForks Books, an imprint of Falcon® Publishing, Inc.,
Helena, Montana.
Printed in Canada.

1 2 3 4 5 6 7 8 9 0 TP 04 03 02 01 00 99

Cover and inside food photographs © 1999 Carol Rublein.
Library of Congress Cataloging-in-Publication Data is on file at
the Library of Congress.

For extra copies of this book and information about other ThreeForks Books,
write to Falcon, P.O. Box 1718, Helena, Montana 59624; or call 1-800-582-2665.
You can also visit our website at www.Falcon.com or contact us by e-mail at
falcon@falcon.com.

# Contents

# -Preface-

Welcome to the wonderful, extended world of Wheat Montana. This *Wheat Montana Cookbook* was compiled using recipes gathered from around Montana and across the country from the many people who have experienced the great products we produce here on our farm. Some recipes come from folks who have discovered us by way of stopping into our deli and bakery while traveling through Montana from afar. Other recipes come from neighbors who live just down the road, yet rely on the same consistent quality of our products as do baking professionals thousands of miles away. Either a family tradition or a newfound discovery, each recipe is a favorite of the person or family who provided it to us. You'll also find some of Wheat Montana's favorite and most unique recipes here.

The only requirement for recipe submission was that the recipes use one or more of the many premium flours and grains available from our farm: Wheat Montana Bronze Chief®, Prairie Gold®, Natural White®, and 7-grain cereal. We have not tested, nor have we altered, the recipes in any way except to note where the different brands of our flours should be used.

We hope you enjoy the great tastes we have compiled here in our very first cookbook. It is as much yours as it is ours since it is our customers who supply the special touch and creativity to make our fine products taste even better.

# –Acknowledgments–

We would like to thank all of you who took the time to respond to our invitation to submit your recipes. Sharing ideas and creativity is what makes the world go around. Without you, this compilation would not be possible.

We would also like to thank Carol Rublein for her beautiful photographs which grace the chapter sections and cover. Her attention to detail is apparent not only in the photos, but also in the time she took to prepare and bake the recipes illustrated herself. We couldn't tell if she was a better chef or a better photographer.

Also, thanks to all of our customers who have offered their encouragement, comments, and letters over the years. You all help us grow this company, and we are grateful.

*The Folkvords*

# -About Wheat Montana-

The Wheat Montana story begins on the family farm. Two special varieties of high-protein wheat are carefully seeded across the 13,000-acre farm of rich prairie soil, between the headwaters of the Missouri River and the Continental Divide. Southwestern Montana's high elevation provides Wheat Montana Farms with excellent growing conditions. Hot summer days and a semi-arid climate ensure optimum moisture levels, while cool nights slow the grain's growth allowing for the full development of size, weight, flavor, and protein content. By harvest time, we are blessed with the finest wheat in the world.

We utilize our wheat in our deli and bakery in Three Forks, Montana, where we produce nearly 10,000 loaves of bread a day. While our bakery churns out a variety of great tasting breads, our deli draws visitors from around the world to sample the fruits of our labor and to take a slice of Montana for themselves. In fact, Wheat Montana products have been praised for quality, taste, and innovation by nationally recognized names such as DuPont, *Food & Wine*, and *Country Living Magazine*.

In addition to our bakery and deli products, we have a rapidly growing national specialty grains business that includes not only our homegrown wheat, but also a large selection of specialty grains and other bulk items. All of these quality items, in addition to Wheat Montana activewear, are available to customers by direct mail or by contacting us at our website: www.wheatmontana.com.

We think you'll agree our commitment to grow the best, provide the best, and bake the best is evident in everything with our name on it.

# Pancakes & Waffles

*Whole Wheat Banana Pancakes*
*(see page 4)*

WHEAT MONTANA
THREE FORKS, MONTANA

---

## GOLDEN WHEAT PANCAKES

1⅓ cups sifted Prairie Gold® whole wheat flour
3 tablespoons sugar
¾ teaspoon salt
3 teaspoons baking powder
3 eggs, well beaten
1¼ cups milk
3 tablespoons shortening, melted

Stir together dry ingredients. Combine eggs and milk, then stir into dry
ingredients along with melted shortening, mixing only until blended. (For lighter
pancakes, eggs may be separated and the white stiffly beaten. Add yolk to
ingredients; fold whites into batter just before baking.) Fry on lightly greased
griddle over low-medium heat until golden brown, then turn. This batter makes
excellent waffles as well.

Makes 12 (4-inch) pancakes.

JEANETTE SOSTROM
ABSAROKEE, MONTANA

## BASIC PANCAKE FLOUR MIX

4 cups Prairie Gold® whole wheat flour
4 cups Natural White® flour
1/3 cup baking powder
1/2 cup sugar
4 teaspoons salt
1 1/4 cups powdered milk

Combine all ingredients thoroughly. Store in an airtight container and use as needed. No need to refrigerate.

Makes about 10 cups of pancake mix.

## PANCAKES FOR TWO

1 1/4 cups basic pancake mix
1 cup water
1 beaten egg
2 tablespoons oil

Mix all ingredients together until smooth. Cook on a hot griddle or skillet until golden brown on both sides.

This is handy to take traveling or camping.

Makes about 4 (5-inch) pancakes.

ESTHER SCHMELING
SENTINEL BUTTE, NORTH DAKOTA

## WHOLE WHEAT BANANA PANCAKES

4 cups Prairie Gold® whole wheat flour
2 teaspoons salt
4 teaspoons baking powder
2 teaspoons baking soda
2 teaspoons cinnamon
2 tablespoons ground flax
4 cups sour milk or buttermilk
½ cup olive oil
1–2 ripe bananas, mashed
4 eggs

Mix all ingredients and fry on griddle until golden brown.

Makes 12–14 plate-sized pancakes.

LENA CLEVELAND
BROADVIEW, MONTANA

## BRONZE CHIEF® WHOLE WHEAT PANCAKES

1 cup Bronze Chief® whole wheat flour
1/2 cup Natural White® flour
2 tablespoons brown sugar
2 teaspoons baking powder
1 teaspoon baking soda
1/2 teaspoon salt
1 1/2 cups buttermilk
2 eggs, separated
2 tablespoons oil

Sift together the flours, brown sugar, baking powder, baking soda, and salt. In a separate bowl, combine buttermilk, egg yolks, and oil. Add to dry ingredients. Beat egg whites until stiff; fold into batter. Fry pancakes on hot griddle and serve with honey or syrup.

Makes 10 pancakes.

LENA CLEVELAND
BROADVIEW, MONTANA

## OVERNIGHT PANCAKES

**Day one:**
2 cups Natural White® flour
1 tablespoon sugar
2 cups buttermilk
1 teaspoon salt
1½ teaspoons baking powder

**Day two:**
2 teaspoons hot water with 1 teaspoon baking soda
1 egg, beaten
1 tablespoon oil

Mix together the flour, sugar, buttermilk, salt, and baking powder and refrigerate the mixture in a covered container overnight.

In the morning, add water-soda mixture, egg, and oil to the refrigerated mixture. Mix gently and fry on hot griddle.

Makes 10 pancakes.

THERESE FAIRBANKS
LIVINGSTON, MONTANA

---

## WHOLE WHEAT WAFFLES

1 cup Natural White® flour
1 cup Prairie Gold® whole wheat flour
1 teaspoon baking soda
½ teaspoon salt
1 tablespoon sugar
2 cups buttermilk
2 eggs
4 tablespoons butter, melted

Sift together the dry ingredients. In separate bowl, combine the buttermilk and eggs and add to the dry ingredients. Add the melted butter and stir to combine. Cook in preheated waffle iron.

Makes about 16 (8-inch) waffles.

# Muffins, Biscuits & Quick Breads

*Blueberry Muffins*
*(see page 20)*

Frances Folkvord, Wheat Montana
Three Forks, Montana

## WHEAT MONTANA 7-GRAIN CARROT MUFFINS

1 cup Wheat Montana 7-grain cereal
½ cup raisins
1 cup skim milk
½ cup shredded carrots
½ cup sugar
½ cup brown sugar
¼ cup vegetable oil
2 eggs
1 teaspoon grated orange peel
1 cup Prairie Gold® whole wheat flour
1 tablespoon baking powder
½ teaspoon baking soda

In a large bowl combine cereal, raisins, and milk. Stir well, cover and refrigerate 2 hours. Preheat oven to 400°. Combine carrots, sugars, oil, eggs, and orange peel. Stir into cereal mixture. Combine dry ingredients and stir into batter until just moistened. Spoon muffin batter into greased muffin tin. Bake in a preheated 400° oven for 20–25 minutes.

Makes 10 muffins.

FRANCES FOLKVORD, WHEAT MONTANA
THREE FORKS, MONTANA

## ZUCCHINI LOAF

2 cups Prairie Gold® whole wheat
1 cup Natural White® flour
1 teaspoon baking soda
1 teaspoon salt
3 teaspoons cinnamon
¼ teaspoon baking powder
1 cup sugar
3 eggs
1 cup vegetable oil
3 teaspoons vanilla
2 cups grated zucchini (remove seeds)
1 cup raisins
½ cup nuts

Preheat oven to 325°.

Sift together flours, baking soda, salt, cinnamon, and baking powder. Add remaining ingredients and mix well. Put bread batter into 2 (4x8-inch) greased and floured loaf pans. Bake 1 hour in 325° oven.

Makes 2 loaves.

DARLEEN ELLINGSON
BOZEMAN, MONTANA

---

## WHEAT GERM MUFFINS

1½ cups canola oil
1 cup brown sugar
2 eggs
1 teaspoon baking powder
1 teaspoon baking soda
1 cup Natural White® flour
1½ teaspoons salt
1 cup sour cream*
1 cup wheat germ

Preheat oven to 400°.

Mix oil and brown sugar; beat in eggs. In a separate bowl, mix baking powder, baking soda, flour, and salt. Add flour mixture to egg mixture alternately with sour cream, adding about a third each addition. Fold in wheat germ. Fill paperlined muffin cups ⅔ full. Bake in a 400° oven for 15–18 minutes.

Makes 12–18 muffins.

*Buttermilk or milk with 2 tablespoons of lemon juice added may be substituted for sour cream.

Darleen Ellingson
Bozeman, Montana

## APPLESAUCE–OATMEAL MUFFINS

1½ cups canola oil
¾ cup brown sugar
1 egg
1¼ teaspoons baking soda
¼ teaspoon salt
1 cup Natural White® flour
½ teaspoon cinnamon
½ cup chopped dates or raisins
¾ cup applesauce
1 teaspoon baking powder
1 cup quick rolled oats
½ cup chopped pecans
Confectioners' sugar for topping

Preheat oven to 350°.

Mix oil and brown sugar until light and fluffy. Add egg and beat well. In a separate bowl, mix baking soda, salt, flour, cinnamon, and dried fruit well; add alternately (a third of each mixture at a time) with applesauce to creamed mixture, stirring well. Add baking powder, oats, and pecans and mix well. Spoon into paperlined muffin cups. Bake in a 350° oven for 25–30 minutes. Cool on rack. Turn out and sift confectioners' sugar over top.

Delicious served with a main dish salad for a luncheon or a simple supper with a casserole.

Makes 12–18 muffins.

Darleen Ellingson
Bozeman, Montana

---

## ZUCCHINI–BASIL MUFFINS

>2 cups Natural White® flour
>1/4 cup sugar
>1 tablespoon baking powder
>1 teaspoon salt
>2 eggs
>3 1/4 cups milk
>2/3 cup canola oil
>2 cups shredded zucchini
>2 tablespoons minced basil
>1/4 cup grated Parmesan cheese

Preheat oven to 425°.

Combine flour, sugar, baking powder, and salt. In a separate bowl, beat eggs; stir in milk and oil. Mix into dry ingredients. Gently mix in zucchini and basil. Fill paperlined muffin cups 2/3 full. Sprinkle with Parmesan cheese. Bake in a 425° oven for 20–25 minutes.

A great accompaniment for a salad supper.

Makes 12–15 muffins.

LINDA BURDIC
ONTARIO, OREGON

## FANCY CORN MUFFINS

1 cup Natural White® flour
1 cup cornmeal
1 tablespoon baking powder
1 teaspoon salt
1⅓ cups fresh Parmesan cheese, finely grated
4 tablespoons butter, softened
½ cup apricot jam
2 eggs
1 (15-ounce) can of creamed corn

Preheat oven to 400°.

In a deep mixing bowl, mix together the first five ingredients. In a smaller bowl, cream together the softened butter and apricot jam. Add eggs and creamed corn to the liquid mixture; mix well. Pour liquid mixture into the flour mixture. Mix gently using a large spoon. Use only 12–14 strokes; do not overmix. Spoon batter into greased muffin tins (or use paperlined muffin cups). Bake for 20–25 minutes or until golden brown color appears on crowns. Serve warm with lots more apricot jam.

Makes approximately 12 large muffins.

LAURA SHERWOOD
KEARNEY, NEBRASKA

---

## WHOLE WHEAT MUFFINS

¾ cup applesauce
¾ cup sugar
2 eggs, beaten
2 cups buttermilk or sour milk
4½ cups Bronze Chief® whole wheat flour
2½ teaspoons baking soda
1 teaspoon salt
Raisins or nuts (optional)

Preheat oven to 400°.

Mix first four ingredients together and then add the rest, mixing slightly. Spray paper muffin cups with non-stick cooking spray and line muffin tin. Spoon batter into paperlined muffin tin.

Bake in a 400° oven for 20 minutes. Batter will keep in the refrigerator for up to 6 weeks.

Makes 24 muffins.

LEILA WILLIAMS
GREAT FALLS, MONTANA

## SUPER GOOD BRAN MUFFINS

2 cups 100% bran cereal (the kind shaped like little worms)
1¼ cups 2% reduced fat milk
1 cup Natural White® flour
1½ teaspoons baking powder
½ teaspoon baking soda
½ teaspoon ground cinnamon (optional)
1 egg
½ cup applesauce
⅓ cup honey
½ cup raisins
¼ cup nuts (optional)

Preheat oven to 400°.

Combine bran and milk in large mixing bowl and let stand 5 minutes. While waiting, in a separate small bowl, mix flour, baking powder, baking soda, and cinnamon. When bran mixture is ready, add the egg, applesauce, honey, raisins, and nuts, stirring them into the bran mixture after each addition. Add the flour mixture to the rest of the batter.

Spoon batter into muffin pan sprayed with non-stick cooking spray, filling each cup ⅔ full. Bake in a 400° oven for 18–20 minutes or until golden brown. Delicious when served warm.

Makes 12 muffins.

ELNA BOHNEN
BILLINGS, MONTANA

---

## RAW APPLE MUFFINS

1 cup sugar
2 eggs, beaten
½ cup vegetable oil or olive oil
2 teaspoons vanilla extract
1½ cups Prairie Gold® whole wheat flour
½ cup Natural White® flour
2 teaspoons baking soda
2 teaspoons cinnamon
1 teaspoon salt
1 cup nuts
4 cups diced, peeled apples
1 cup raisins

Preheat oven to 325°.

Mix all ingredients well; add apples and raisins last. Bake in a 325° oven for 30 minutes or more until golden brown.

Makes 12 muffins.

THERESA LODE
HELENA, MONTANA

## PUMPKIN BRAN MUFFINS

1 cup wheat bran
⅓ cup brown sugar
1¼ cups Prairie Gold® whole wheat flour
2 teaspoons baking powder
½ teaspoon baking soda
1 teaspoon cinnamon
1 teaspoon ground ginger
1 cup raisins
3 large egg whites
1 cup pumpkin
¾ cup buttermilk or yogurt (nonfat)
⅓ cup molasses

Preheat oven to 500° then reduce to 400°.

Whisk first 7 ingredients together, then add raisins. In a separate bowl, mix remaining ingredients including eggs, then combine with flour mixture just enough to moisten. Do not overmix.

Spray muffin tins with non-stick cooking spray. Fill muffin tins ⅔ full. Bake for about 25 minutes in a 400° oven. Let cool a few minutes before removing from tin.

Makes about 12 muffins.

CARLEEN EISELE
DIXON, ILLINOIS

---

## BLUEBERRY MUFFINS

2 eggs, beaten
1 cup milk
½ cup canola oil
3 cups Natural White® flour
1 cup sugar
4 teaspoons baking powder
1 teaspoon salt
2 cups blueberries, fresh or drained canned

Preheat oven to 400°.

Blend eggs, milk, and oil in a large bowl. In a separate bowl, combine flour, sugar, baking powder, and salt. Gradually add flour mixture to liquids. Beat well. Fold in blueberries. Divide batter evenly into paperlined muffin cups or a muffin tin coated with non-stick cooking spray. Bake in a 400° oven for 20 minutes.

Makes 12 muffins.

CAROL GARLOW
LIVINGSTON, MONTANA

## BANANA–STRAWBERRY MUFFINS

3 cups Natural White® flour
1½ cups sugar
1 teaspoon baking powder
½ teaspoon salt
4½ teaspoons ground cinnamon
1¼ cups milk
2 eggs
1 cup (2 sticks) unsalted butter, melted
3 medium-sized bananas, mashed
½ cup diced fresh or frozen strawberries

Preheat oven to 375°.

Place paper liners in 24 muffin cups. In large bowl, combine and mix flour, sugar, baking powder, salt, and cinnamon. Add milk, eggs, and melted butter into dry mixture until well blended. Fold in bananas and strawberries. Spoon batter into muffin cups. Bake in a 375° for 20 minutes or until golden brown.

Makes 24 muffins.

JOYCE SUTTON
LIVINGSTON, MONTANA

## 7-GRAIN MUFFINS

> 1 cup Wheat Montana 7-grain cereal
> 1 cup sour cream or buttermilk
> ⅓ cup cooking oil
> ½ cup brown sugar
> 1 egg
> 1 cup Bronze Chief® whole wheat flour
> 1 teaspoon baking powder
> ½ teaspoon baking soda
> 1 teaspoon salt

Soak the 7-grain cereal in the sour cream for 1 hour.

Preheat oven to 400°.

Mix the oil, brown sugar, and egg together in a large mixing bowl. Sift flour, baking powder, baking soda, and salt in separate bowl. Stir the flour mixture into the sugar mixture alternately with the 7-grain mixture.

Fill greased muffin pan cups ⅔ full. Bake in a 400° oven until golden brown. Serve hot.

Makes 12 muffins.

Deena Gobbs
Townsend, Montana

## ALL-BRAN MUFFINS

1 cup All-Bran
¾ cup milk
1 cup Bronze Chief® whole wheat flour
½ teaspoon salt
2 teaspoons baking powder
½ cup sugar
2 tablespoons oil
1 egg
½ cup raisins (optional)

Preheat oven to 400°.

Combine All-Bran and milk. Mix flour, salt, and baking powder together. In a separate bowl, mix sugar, oil, and egg, beating briskly for 1 minute. Add flour mixture and beat until smooth. Add raisins. Spoon batter into 12-cup muffin tin and bake in a 400° oven for 15–20 minutes.

Makes 12 muffins.

DEENA GOBBS
TOWNSEND, MONTANA

## WHOLE WHEAT DINNER MUFFINS

1 cup unsifted Natural White® flour
1 cup unsifted Bronze Chief® whole wheat flour
1/4 cup wheat germ
3 teaspoons baking powder
1/2 teaspoon salt
1/4 cup sugar, brown or white, or 2 tablespoons honey
1 egg
4 tablespoons margarine or butter, melted, or oil
1 cup milk

Preheat oven to 375°.

In a large bowl, stir dry ingredients together (including sugar if using brown or white). In separate bowl, beat egg and stir in shortening, sweetener (if using honey), and milk. Mix dry and wet ingredients together to moisten. Spoon batter into greased muffin cups, 2/3 full. Bake in a 375° oven for about 25 minutes.

Makes about 12 muffins.

Darleen Ellingson
Bozeman, Montana

## DAR'S BISCUITS

2 cups Natural White® flour
3 teaspoons baking powder
1 teaspoon salt
⅓ cup canola oil
⅔ cup milk

Preheat oven to 475°.

Measure dry ingredients into bowl. Pour oil and milk into measuring cup, but do not stir. Pour all at once into dry mixture. Stir with a fork until mixture cleans sides of bowl and rounds into a ball. Smooth by kneading dough about 10 times without additional flour. Gently roll to about ¼-inch thickness. Cut into rounds. Bake in a 475° oven for 6–8 minutes on ungreased baking sheet.

Makes about 16 biscuits.

DEON TOMSHECK
ETHRIDGE, MONTANA

---

## BISCUITS SUPREME

2 cups sifted Natural White® flour
4 teaspoons baking powder
½ teaspoon cream of tartar
½ teaspoon salt
2 teaspoons sugar
½ cup shortening
⅔ cup milk

Preheat oven to 450°.

Sift dry ingredients into bowl. Cut in shortening until crumbly. Stir in milk until mixture forms a ball. Dough will be soft. Turn onto lightly floured surface and knead gently 10–12 times. Pat or roll dough ½-inch thick. Cut with floured biscuit cutter; do not twist cutter. Place rounds on ungreased baking sheet. Bake in a 450° oven for 10–12 minutes.

Makes 12 biscuits.

Velda Hensley Welch
Toston, Montana

## QUICK-MIX BISCUIT PLUS MIX

8½ cups Natural White® flour (up to ½ can be Prairie Gold®
   whole wheat flour)
1 tablespoon baking powder
1 tablespoon salt
1 tablespoon baking soda
2 teaspoons cream of tartar
1½ cups instant dry milk
2¼ cups shortening

In a large bowl, sift together all dry ingredients. Blend well. With pastry blender cut in shortening. Put in a large, airtight container. Store in a cool dry place. Use within 11 weeks.

Makes 13 cups.

This mix can be used like Bisquick mix or other biscuit mix. Use enough liquid to make desired amount of biscuits and bake, following general biscuit baking directions.

Melvena Hartford
Lewistown, Montana

## PUMPKIN BREAD

2 eggs
2 teaspoons vanilla
3/4 cup (1 1/2 sticks) butter
1 1/2 cups white granulated sugar
1 1/2 cups brown sugar
2 1/2 cups Prairie Gold® whole wheat flour
2 1/2 cups Natural White® flour
3 teaspoons baking soda
1/2 teaspoon baking powder
3/4 teaspoon salt
1 1/2 teaspoons cinnamon
1 1/2 teaspoons cloves
2 cups raisins
1 cup chopped walnuts
1 cup evaporated milk
1 large can Libbys pumpkin
1/2 cup water

Preheat oven to 350°.

Using a mixer, combine eggs, vanilla, butter, and sugars. In a separate bowl, sift together flours, baking soda, baking powder, salt, cinnamon, cloves, raisins, and walnuts. Slowly add the sifted dry ingredients alternately with evaporated milk, pumpkin, and water to the egg mixture. Grease and flour 4–5 (8x4x2-inch) loaf pans. Fill each pan 2/3 full.

Bake in a 350° oven for 1 hour if using small pans. Bake 90 minutes if using larger pans. Check loaves for doneness frequently.

Makes 4–5 small loaves.

THERESA LODE
HELENA, MONTANA

## ZUCCHINI BREAD

1¾ cups Prairie Gold® whole wheat flour
⅔ cup oat bran
½ cup sugar
1 tablespoon dry milk
2 teaspoons baking powder
½ teaspoon cinnamon
1 teaspoon baking soda
½ teaspoon salt
2 large eggs (or 3 large egg whites)
1 cup buttermilk
1½ cups shredded or puréed zucchini

Preheat oven to 425°. Grease a 9x5-inch loaf pan.

Whisk all the dry ingredients in large bowl. In a separate bowl, lightly beat eggs. Add buttermilk and zucchini. Add liquid ingredients to the flour mixture and combine. Don't overbeat! Pour dough into greased pan. Bake in a 425° oven for 40 minutes.

Makes 1 loaf.

VALERIE MIDDLEMAS
TOWNSEND, MONTANA

## CANADIAN BANANA BREAD

1 cup brown sugar
4 tablespoons margarine
1 teaspoon vanilla
3 large, ripe bananas, mashed
2 eggs, well beaten
2 cups Prairie Gold® whole wheat flour
3 teaspoons baking powder
½ teaspoon salt

Preheat oven to 325°.

Cream sugar and margarine, then add in turn, vanilla, bananas, and eggs. In a separate bowl, mix flour, baking powder, and salt. Add flour mixture to banana mixture. Divide batter evenly into 2 greased loaf pans. Bake in a 325° oven for 1 hour.

Makes 2 loaves.

CARLEEN EISELE
DIXON, ILLINOIS

## BANANA BREAD

2 tablespoons brown sugar
1/2 cup (1 stick) butter, softened
1 cup sugar
2 eggs, unbeaten
3 ripe bananas, mashed
1 tablespoon sour milk or buttermilk
1/2 teaspoon vanilla
1 1/2 cups Natural White® flour
1/2 cup Bronze Chief® whole wheat flour
1 teaspoon salt
1 teaspoon baking soda

Preheat oven to 350°.

Cream together the brown sugar, butter, and sugar. Add the eggs, bananas, sour milk, and vanilla, and blend well. In a separate bowl, combine flours, salt, and baking soda. Combine all ingredients to make batter. Stir until blended. Pour evenly into 2 lightly oiled loaf pans. Bake in a 350° oven for 1 hour. Remove immediately from pans, cool on wire rack.

Makes 2 loaves.

Deena Gobbs
Townsend, Montana

## WHOLE WHEAT BANANA BREAD

½ cup (1 stick) butter or margarine
1 cup sugar
2 eggs, lightly beaten
1 cup mashed banana
1 cup Natural White® flour, unsifted
½ teaspoon salt
1 teaspoon baking soda
1 cup Prairie Gold® or Bronze Chief® whole wheat flour,
    unsifted
⅓ cup hot water
½ cup chopped nuts

Preheat to oven 325°.

Melt butter and blend in sugar. Mix in beaten eggs and mashed banana, and blend until smooth. Stir together white flour, salt, baking soda, and whole wheat flour. Add dry ingredients alternately with hot water to banana mixture. Stir in chopped nuts. Spoon batter into greased 9x5-inch loaf pan. Bake in a 325° oven for 1 hour and 10 minutes or until done. Cool in pan for 10 minutes, then turn on rack to finish cooling.

Makes 1 loaf.

BETTY HORNE
TOWNSEND, MONTANA

## APPLESAUCE BROWN BREAD

2 cups Prairie Gold® whole wheat flour
1 cup cornmeal
¾ teaspoon salt
1 teaspoon baking soda
1 cup buttermilk
1 cup dark molasses
¾ cup sweetened applesauce
¾ cup raisins

Preheat to oven 350°.

Combine dry ingredients in large mixing bowl. Add buttermilk and molasses, and beat until smooth. Mixture will be thick. Fold in applesauce and raisins. Turn into greased 9-inch square baking pan and bake in a 350° oven for 35 minutes or until done. Cool 10 minutes in pan, then remove and cool on wire rack. Cut into squares.

Makes 16 squares.

ELAINE HENSLEY
TOSTON, MONTANA

## PRAIRIE BISCUITS

2 cups Natural White® flour
½ cup lard
1 teaspoon salt, or less to taste
½ teaspoon baking soda
4 teaspoons baking powder
¾ cup buttermilk

In bag of flour, make a "well." Mix in lard (crumble the lard and flour with your hands) and work in the salt, baking soda, and baking powder. Sprinkle the buttermilk into dough. Form a ball, knead on a floured surface or in hands. Form dough into size of biscuits desired and bake in a hot dutch oven over coals, or in a preheated 450° oven until tannish in color.

Number of biscuits depends on size formed.

BETTY HORNE
TOWNSEND, MONTANA

## STEAMED BROWN BREAD

1 cup Natural White® flour
1 cup Bronze Chief® whole wheat flour
1 cup cornmeal
1½ teaspoons salt
½ cup sugar
1 teaspoon baking soda
½ cup molasses
1½ cups buttermilk
2 tablespoons shortening, melted

Sift flours, cornmeal, salt, and sugar together. In a separate bowl, mix baking soda and molasses. Add buttermilk and molasses mixture to the dry mixture. Add shortening. Mix well. Fill a greased 2½-quart pudding mold ¾ full and steam for 3 hours. Remove cover from pan and allow to cool 20 minutes. Remove to wire rack and finish cooling.

Makes 1 loaf.

ADAH HORNE
TOWNSEND, MONTANA

## QUICK OATMEAL BREAD

1½ cups quick rolled oats
1 cup Prairie Gold® whole wheat flour
1 envelope Spiced Apple Cider mix
2 packets sugar substitute
1 teaspoon baking soda
1 cup buttermilk
1 cup hot water

Preheat oven to 350°

Mix all dry ingredients together. Add buttermilk, then hot water. Spoon into greased loaf pan and bake in a 350° oven for 25–30 minutes. Raisins or dates can be added, or sprinkle cinnamon on top before baking. Keep in refrigerator.

Makes 1 loaf.

# Yeast Breads

*Buttermilk Wheat Rolls*
*(see page 52)*

WHEAT MONTANA
THREE FORKS, MONTANA

## WHEAT MONTANA WHITE BREAD

2 tablespoons yeast
2¼ cups warm water (105°)
7 cups Natural White® flour
4 tablespoons honey
1 tablespoon salt
3 tablespoons shortening
3 tablespoons powdered milk

Dissolve yeast in warm water and let stand 5 minutes. Add flour, honey, salt, shortening, and powdered milk. Mix until well developed. Dough should be able to be stretched into a thin, transparent window.

Place dough in oiled bowl and let double in size. Punch down and divide into 2 equal pieces. Flatten each piece and roll into bread loaves. Place into bread pans and let rise until above sides of pan.

Bake in preheated 350° oven for 25–35 minutes, until golden on top. Bread should sound hollow when tapped.

Makes 2 (1½-pound) loaves.

WHEAT MONTANA
THREE FORKS, MONTANA

## WHEAT MONTANA WHEAT BREAD

2 tablespoons yeast
2½ cups warm water (105°)
3½ cups Natural White® flour
3½ cups Bronze Chief® whole wheat flour
4 tablespoons honey
1 tablespoon salt
3 tablespoons shortening
3 tablespoons powdered milk

Dissolve yeast in warm water and let stand 5 minutes. Add flour, honey, salt, shortening, and powdered milk. Mix until well developed. Dough should be able to be stretched into a thin, transparent window.

Place dough in oiled bowl and let double in size. Punch down and divide into 2 equal pieces. Flatten each piece and roll into bread loaf. Place into bread pans and let rise until above sides of pan.

Bake in preheated 350° oven for 25–30 minutes, until golden on top. Bread should sound hollow when tapped.

Makes 2 (1½-pound) loaves.

Frances Folkvord, Wheat Montana
Three Forks, Montana

---

## 7-GRAIN WHOLE WHEAT BREAD FOR BREAD MACHINE

½ cup Wheat Montana 7-grain cereal
2 cups warm water
1½ cups skim milk
2 tablespoons butter
2 cups Natural White® flour
1 cup Bronze Chief® whole wheat flour
3 tablespoons honey
1¼ teaspoons salt
2 teaspoons active dry yeast

Soak 7-grain cereal in warm water for several hours. Drain well. Carefully measure all ingredients into pan and follow machine instructions.

Makes 1 loaf.

## Wheat Montana
## Three Forks, Montana

---

# DELICIOUS WHITE BREAD

2 packages active dry yeast
2¼ cups lukewarm water
½ cup nonfat dry milk
2 tablespoons sugar
1 tablespoon salt
⅓ cup cooking oil
7–7½ cups sifted Natural White® flour

Sprinkle yeast on lukewarm water; stir to dissolve. Add dry milk, sugar, salt, oil, and 3 cups flour. Beat with electric mixer at medium speed until smooth (about 3 minutes), scraping bowl occasionally. (Or beat with spoon until batter is smooth.) Gradually add enough remaining flour to make a soft dough that cleaves to the sides of the bowl. Cover; let rest for 15 minutes.

Knead on a floured surface until smooth and satiny, about 5 minutes. Divide dough in half; shape each half into a loaf. Let rise until doubled, about 1–1½ hours. Bake in a preheated 400° oven for 30–35 minutes or until loaves sound hollow when tapped. Remove from pans; cool on racks.

Makes 2 loaves.

WHEAT MONTANA
THREE FORKS, MONTANA

## WHEAT MONTANA FRENCH BREAD

1 tablespoon yeast
1¼ cups warm water (105°)
3 cups Natural White® flour
1  tablespoon sugar
2 teaspoons salt

**For extra flavor, add:**
2 tablespoons Italian seasoning
¼ teaspoon garlic powder or
¼ cup sun-dried tomatoes

Dissolve yeast in warm water and let stand 5 minutes. Add flour, sugar, and salt. Add extra flavors if using. Mix dough until developed. Dough should be smooth in appearance.

Place dough in oiled bowl and let double in size. Punch down and roll into a long or round loaf. May be divided into 6 equal pieces for hard rolls. Place on baking sheet and let double in size.

Brush with cold water. Bake in a preheated 400° oven for 25–30 minutes. Bread should sound hollow when tapped.

Makes 1 (1½-pound) loaf or 6 (4-ounce) rolls.

## WHEAT MONTANA
## THREE FORKS, MONTANA

## WHEAT MONTANA BUN DOUGH

> 2 tablespoons yeast
> 1 cup warm water (105°)
> 3½ cups Natural White® flour
> ¼ cup sugar
> ¼ cup shortening
> 1 teaspoon salt
> 3 tablespoons powdered milk
> 1 egg, for a lighter texture (optional)

Dissolve yeast in warm water and let stand 5 minutes. Add flour, sugar, shortening, salt, powdered milk, and egg. Mix dough until well developed. Dough should be able to be stretched into a thin, transparent window.

Place dough in oiled bowl and let double in size. Punch down and divide in 12 equal pieces for hamburger buns or 24 pieces for dinner rolls. Using palm of hand, round each dough piece and place on baking sheet. Let rise until doubled in size.

Bake in a preheated 375° oven for 12–18 minutes, until golden in color.

WHEAT MONTANA
THREE FORKS, MONTANA

---

## TRADITIONAL WHOLE WHEAT BREAD

1 package active dry yeast
⅓ cup lukewarm water
1 tablespoon shortening, melted
1 tablespoon honey
1 tablespoon molasses
1 tablespoon salt
3 cups scalded milk
6 cups Bronze Chief® whole wheat flour

Soften yeast in water. Combine melted shortening, honey, molasses, salt, and scalded milk. Let cool to a lukewarm temperature and combine with yeast mixture. Add enough flour as needed, until it can be handled without sticking to your hands. Shape into 2 loaves and place in greased loaf 4x8-inch tins. Let rise to not quite double in bulk (1–2 hours). Bake in a preheated 350° for 1 hour and 10 minutes.

Makes 2 loaves.

## WHEAT MONTANA
## THREE FORKS, MONTANA

### PERFECT BREAD MACHINE RECIPE

1⅛ cups water

2 cups Prairie Gold® or Bronze Chief® whole wheat flour

1½ tablespoons sugar

1¼ teaspoons salt

1½ tablespoons butter

1½ tablespoons dry milk

1 envelope active dry yeast

Pour water into the bread pan. Add flour, sugar, salt, butter, and dry milk. Hollow out the center of the dry ingredients and put in yeast. If yeast contacts the water before kneading, the bread may not rise well. All ingredients should be at room temperature (70–80°). When room temperature is below 65°, use lukewarm water (about 100°). Activate your bread machine following its instructions.

*Variations to instructions may apply based on your bread machine model. Refer to your machine manual.*

MARILYN STEINGRUBER
MANHATTAN, MONTANA

## BACON–ONION–OATMEAL BUNS

>    2 cups boiling water
>    1 cup rolled oats
>    2 packages granular yeast
>    ⅓ cup warm water for yeast
>    3 tablespoons Crisco oil
>    ¼ cup dark molasses
>    ⅔ cup brown sugar
>    1 egg
>    2 teaspoons salt
>    5–6 cups Natural White® flour
>    ¾ pound bacon cut into ¼-inch strips (fried not too crisp)
>    2 cups diced onion (fried in bacon drippings until slightly
>       brown)

Pour boiling water over oats. Cool. Dissolve yeast in water. To oatmeal, add dissolved yeast, oil, molasses, brown sugar, egg, and salt. Beat in enough flour to make a soft dough. Add bacon, onion, and remaining flour. Knead and let rise until doubled in size; punch down and let rise again. Roll the dough into a long log shape and slice into 30 equal bun-sized pieces. Place buns in a buttered pan. Let rise until doubled in size.

Bake in a preheated 375° oven for 18–20 minutes. Brush the buns with melted butter as you take them out of the oven.

Makes 26–30 buns.

DEBBY HANSMANN
HELENA, MONTANA

## GREATMA'S BUNS

3 cups milk
3 cups boiling water
1⅓ cups oil
9 teaspoons yeast
1¾ cups sugar
4 eggs, beaten
4 teaspoons salt
16 cups Natural White® flour

In a large mixer bowl, combine milk, water, and oil. Add remaining ingredients. Mix but do not knead. Let the dough rise until doubled in size, punch it down, then let rise until doubled again. Form dough into 2-inch buns and place on a greased baking sheet. Let rise until doubled in size.

Bake in a preheated 350° oven for 10–12 minutes.

Makes 7 dozen buns.

CAROL GARLOW
LIVINGSTON, MONTANA

## FOCACCIA

### Dough:
1¾ cups Natural White® flour
1¼-ounce package active dry yeast
1 teaspoon sugar
¾ teaspoon salt
¾ cup hot water
2½ tablespoons olive oil

### Topping:
1½ tablespoons olive oil
½ cup pesto
½ cup crumbled feta cheese
6 sundried tomatoes, diced

In a large bowl, combine flour, yeast, sugar, and salt. In a small bowl, combine water and olive oil. Slowly add to flour mixture, stirring to form sticky dough. Turn dough onto lightly floured surface; knead in additional flour as needed until dough is smooth and elastic. Place dough in oiled bowl, turning to coat entire surface. Cover and let rise until doubled in size, about 40 minutes.

Grease 13-inch round baking sheet. Punch down dough and let rise 5 minutes. Turn dough onto lightly floured surface. Using floured rolling pin, roll dough into 12-inch round. Place onto prepared baking sheet. Build up edges of dough to form crust. Cover and let rise 15–30 minutes. Drizzle with olive oil; add pesto, feta, and tomatoes. Bake in a preheated 375° oven for 30 minutes or until lightly browned on top and sides. Remove from oven, cool, and cut into wedges. Serve warm.

Serves 4–6.

DEENA GOBBS
TOWNSEND, MONTANA

## WHOLE WHEAT ROLLS

2 tablespoons yeast
1 teaspoon sugar
½ cup warm water
2 eggs, beaten
½ cup oil
½ cup honey
1 teaspoon salt
½ cup canned condensed milk
½ cup hot water
3–4 cups Prairie Gold® or Bronze Chief® whole wheat flour

Soften yeast and sugar in ½ cup warm water. Mix together eggs, oil, honey, salt, milk, and ½ cup hot water. Add yeast mixture. Add flour; mix all. Let dough rest for 10–15 minutes. Knead dough, let rise. Punch down, knead again, and let rise until doubled in bulk. Make into desired size dinner rolls.

Bake in a preheated 350° oven for 20–25 minutes.

LINDA BURDIC
ONTARIO, OREGON

## SPROUTED WHEAT BREAD

$\frac{1}{2}$ cup Prairie Gold® or Bronze Chief® wheat berries
1 cup cold water
3 cups warm water (110°)
2 tablespoons active dry yeast
$1\frac{1}{2}$ teaspoons salt
$\frac{1}{3}$ cup granulated cane sugar (may substitute honey, sugar, brown sugar, etc.)
$\frac{1}{3}$ cup olive oil
1 cup wheat germ
4 cups Bronze Chief® whole wheat flour
$3\frac{1}{2}$–4 cups Natural White® flour

**2–3 days before baking bread:**
Soak wheat berries in 1 cup of cold water for 6–8 hours in a sprouting jar. Drain well and rinse several times. Invert or place jar on its side in a warm location (70–80°). Water seeds 3–4 times a day by covering with tap water and then draining well. Your sprouts are ready when they have a $\frac{1}{4}$-inch top and 1-inch roots (about 2 cups).

**Baking day:**
Place sprouts in a blender with 2 cups of warm water and blend on high until the water looks very milky. Strain this milky water into your bread bowl, reserving the sprouts. Add remaining cup of warm water to the bowl. Add yeast to the liquid and let stand until bubbly. Add salt, sugar, oil, and sprouts to the liquid and begin mixing by hand or bread hook. Add wheat germ, mix well. Add 1 cup of whole wheat flour at a time, mixing well after each addition. Finish bread dough by adding white flour 1 cup at a time until mixture can be turned onto a bread board and finished by hand-kneading. When the dough is elastic and smooth, place in an oiled bowl and cover. Let rise 1–$1\frac{1}{2}$ hours or until doubled in size.

Punch down and shape into 3 loaves. Place in greased bread pans and cover.

When loaves have doubled, about 30 minutes, bake in a preheated 350° oven for 1 hour or until tops are nicely browned and the loaves sound hollow when tapped. Remove baked loaves from their pans and cool completely on wire racks before storing in plastic bags. This bread will keep longer if stored in the refrigerator.

Makes 2 loaves.

BETTY HORNE
TOWNSEND, MONTANA

## BUTTERMILK WHEAT ROLLS

1 package active dry yeast
¼ cup warm water (110°)
1¾ cups buttermilk (room temperature)
2 eggs, lightly beaten
⅓ cup firmly packed brown sugar
1½ teaspoons salt
4 tablespoons butter or margarine, melted and cooled
3 cups Prairie Gold® or Bronze Chief® whole wheat flour,
    unsifted

 3½–4 cups Natural White® flour, unsifted

In a large bowl, dissolve yeast in warm water. Add buttermilk, eggs, brown sugar, salt, and butter. Gradually add wheat flour and 1 cup of white flour, and beat 5 minutes. Then gradually mix in enough of remaining white flour to make a stiff dough. Turn out onto floured board, knead until smooth, 10–20 minutes, adding flour as needed to prevent sticking. Turn dough over in a greased bowl, cover, and let rise in warm place until doubled in size, about 2 hours. Punch down dough, knead briefly. Shape as desired.

For rolls, let rise and bake in a preheated 375° oven for 20 minutes or more, until golden brown. For loaves, let rise and bake in a preheated 350° oven for about 30 minutes.

Makes 20–24 rolls or 2 loaves.

LENA CLEVELAND
BROADVIEW, MONTANA

## PILGRIM BREAD

2 tablespoons active dry yeast
2 cups *plus* 1/2 cup warm water (110°)
1/2 teaspoon sugar
1/2 cup cornmeal
1/3 cup brown sugar
1/4 cup oil
1 1/4 cups Prairie Gold® whole wheat flour
3/4 cup Bronze Chief® whole wheat flour
1/4 cup wheat germ
5 1/2–6 cups Natural White® flour

Dissolve yeast in 1/2 cup warm water. Sprinkle sugar over yeast mixture. Bring remaining 2 cups water to a boil, then pour over cornmeal and brown sugar, and let cool. Mix together yeast mixture, cornmeal mixture, oil, wheat flours, and wheat germ. Beat well, about 2–3 minutes. Add the white flour, kneading until dough is smooth and elastic, 8–10 minutes. Place in a greased bowl, turn to grease top. Let rise in warm place till doubled in size, about 1 1/2 hours.

Punch down, shape into 3 large loaves loaves, let rise in pans till doubled in size. Bake in a preheated 375° oven for 35–45 minutes.

Makes 3 large loaves.

MARION KLAUS
SHERIDAN, WYOMING

## WHOLE WHEAT SOURDOUGH BREAD

1 teaspoon active dry yeast
½ cup wheat germ
1 cup Natural White® flour
1 cup Bronze Chief® whole wheat flour
2 tablespoons olive oil
1 tablespoon molasses
Dash iodized salt
⅔ cup skim milk or buttermilk
1 cup sourdough starter (recipe below)
½ cup raw sunflower seeds (optional)

Use this recipe in your bread machine.

**Sourdough starter:** Make sourdough starter by combining 2 cups milk, 2 cups Bronze Chief® whole wheat flour, and 2½ teaspoons yeast. Let sit for a week and stir occasionally. Keep in the refrigerator between uses. (After using starter in recipe, replenish with equal amounts of flour and milk.)

Add the ingredients to the bread machine in the order given. Set the machine on whole wheat dough.

Turn the dough out on a floured board and shape into a medium-sized loaf pan. Follow bread machine directions.

Bake for 30 minutes in a preheated 350° oven.

Makes 1 medium-sized loaf.

LAURIE BARNARD
RED LODGE, MONTANA

## BREAD MACHINE BREAD

1¼ cups warm water (less ½ tablespoon for high elevations)
1 tablespoon honey
1 tablespoon molasses
2 tablespoons soft butter (cut in small pieces)
1 teaspoon salt
1 cup Prairie Gold® whole wheat flour
2¼ cups Natural White® flour
¼ cup gluten flour
2 teaspoons active dry yeast (less ¼ teaspoon for high elevations)

Put water, honey, and molasses in bread machine container. Add butter pieces to liquid. Add remaining ingredients in order, making sure to keep yeast away from liquid ingredients at beginning. Start bread machine on regular bread cycle.

Makes 1 loaf.

DEENA GOBBS
TOWNSEND, MONTANA

## BASIC SWEET DOUGH—WHOLE WHEAT

2 tablespoons active dry yeast
1/2 cup warm water
1 teaspoon sugar
2 eggs, beaten
1/2 cup honey
1 teaspoon salt
1/2 cup lard
1/2 cup condensed milk or cream
1/2 cup hot water
31/2–4 cups Prairie Gold® whole wheat flour

Soften yeast in 1/2 cup warm water and sugar mixed. Combine beaten eggs, honey, salt, lard, milk, and 1/2 cup hot water. Stir in yeast mixture. Add flour a small amount at a time, beating well. Add only enough flour to make a soft dough. Cover and let rise 10 minutes. Turn onto floured board and knead well. Place dough in a greased bowl. Cover tightly and let rise until doubled in size, about 1 hour. Punch down and let rise 30 minutes; punch down, then let dough rest for a final 10 minutes.

Bake in a preheated 350° oven for 20-25 minutes.

Use this dough to make great cinnamon rolls.

Makes about 2 dozen rolls.

Valerie Middlemas
Townsend, Montana

## OVERNIGHT BUNS

1 package active dry yeast
½ cup warm water
4 cups hot water
3 tablespoons shortening
1½ cups sugar
4 teaspoons salt
3 eggs, beaten
6 cups Bronze Chief® or Prairie Gold® whole wheat flour
6 cups Natural White® flour

Allow 5 hours for preparation prior to overnight rising.

Dissolve yeast in ½ cup warm water. Have ready in large bowl: 4 cups hot water, shortening, sugar, salt, and eggs. Add yeast when lukewarm. Add flours to make a bread-like dough. Knead dough well and return to greased bowl. Punch down every ½ hour for approximately 4½ hours. Form dough into egg-sized buns. Cover with a damp cloth and let sit in cold oven overnight.

In the morning, bake in a preheated 400° oven for 15–18 minutes until golden brown. Freezes well.

Makes about 8 dozen buns.

RICHARD WEVLEY
CUT BANK, MONTANA

## SWEET BUNS (ZWIEBACK BREAD)

1 teaspoon *plus* 1 cup sugar
1/2 cup lukewarm water
2 packages active dry yeast
3 cups skim milk
2/3 cup lard, melted
1 tablespoon salt
2 eggs, lightly beaten
81/3–81/2 cups Natural White® flour

Dissolve 1 teaspoon sugar in lukewarm water. Sprinkle yeast over liquid and let rise in warm place for 10 minutes until foaming.

Scald milk, remove from heat, and add melted lard, 1 cup sugar, and salt, stirring to dissolve. When cooled slightly, add eggs and yeast mixture to milk mixture. Slowly add the flour, adding more as needed to keep the dough from being too sticky. Shape dough into a ball and place in a lightly greased bowl. Cover with plastic wrap or towel. Let rise in a warm place for about 1 hour.

Punch down dough and knead for 2 minutes. Let rise again until doubled in size.

Pinch off balls of dough about the size of an egg. Place on a greased baking sheet 1½ inches apart. Let rise 30 minutes. Preheat oven to 375°.

Bake buns in a 375° oven for 15–20 minutes, until golden brown.

Makes about 5 dozen buns.

JOYCE SUTTON
LIVINGSTON, MONTANA

## MOLASSES WHEAT ROLLS

2 packages active dry yeast
½ cup lukewarm water
1½ cups milk, scalded
⅔ cup molasses
2 teaspoons salt
2 eggs, unbeaten
3¾ cups Natural White® flour
3 cups unsifted Bronze Chief® or Prairie Gold®
   whole wheat flour
⅓ cup cooking oil

Dissolve yeast in warm water as directed on package. Cool milk to lukewarm and add to yeast. Stir in molasses, salt, and eggs. Combine flours and add 4 cups of flour mixture to milk mixture. Beat until smooth. Beat in cooking oil, and add remaining flour gradually.

Place dough on floured board. Knead at least 2 minutes, then place in large, greased mixing bowl. Turn dough to grease all sides, cover, and let rise in warm place, out of draft, for 1 hour or until doubled in size. Knead to get air out, grease top lightly, cover, let rise again until doubled in size, approximately 1 hour.

Punch down dough and knead. Dough can be divided into 2 loaves or broken into 24 dinner rolls. Place shaped dough in greased pans. Cover and let rise until doubled in size, approximately 1 hour.

For loaves, bake in a preheated 400° oven for 15 minutes; reduce heat to 350° and bake 30 minutes longer. Remove from pans immediately. Cool on wire racks. Bake rolls in a preheated 350° oven for 25–30 minutes.

Makes 2 loaves or 24 dinner rolls.

JERRY D. WHITMER
BILLINGS, MONTANA

## BREAD MACHINE WHOLE WHEAT BREAD

1 cup plus 2½–3 tablespoons warm water (105°)
1 tablespoon vegetable oil
1½ tablespoons honey
1 tablespoon molasses
1 fresh large egg
½ teaspoon iodized salt
2 cups Natural White® flour (rounded top)
½ teaspoon bread machine yeast

Add first 6 ingredients in order to bread machine basket. Add flour on top of all ingredients to form a rounded top. Form a small well in the center of the flour and add yeast. Stir gently in basket to mix before starting the cycle. Start machine on regular bread cycle, adding ½ teaspoon water if mixture is dry.

Makes 1 (1½-pound) loaf.

### VARIATIONS:

Add ½ cup raisins after machine mixes for 3-5 minutes, or add about 4 tablespoons of unsalted sunflower seeds.

Jeanette Sostrom
Absarokee, Montana

## SESAME SUNFLOWER BREAD

3⅛ cups water
⅓ cup oil
1½ teaspoons salt
3 tablespoons sugar
2 cups *plus* 1–2 cups Natural White® flour
2 cups *plus* 3 cups Bronze Chief® whole wheat flour
3 tablespoons active dry yeast
2 eggs
⅓ cup sesame seeds
⅓ cup raw sunflower seeds

Combine water, oil, salt, and sugar in a microwave-safe bowl. Warm the mixture in the microwave for 2 minutes. Pour heated mixture into a large bowl and stir in 2 cups of each of the flours and yeast. Mix well by hand or using an electric mixer. Add eggs, sesame seeds, sunflower seeds, remaining 3 cups of wheat flour and remaining 1–2 cups white flour. Knead 10 minutes, place in greased bowl, let rise until doubled in size. (This dough rises rather quickly.)

Divide dough into 3 parts and form into loaves. Place into 3 greased 4½x8½-inch pans. Let rise until doubled in size.

Preheat oven to 375°. Bake for 30 minutes at 375°.

Makes 3 loaves.

Laura Sherwood
Kearney, Nebraska

## 7-GRAIN WHOLE WHEAT BREAD

    1 cup Wheat Montana 7-grain cereal
    2 cups *plus* 3 cups warm water (105–115°)
    2 packages active dry yeast
    ⅔ cup honey
    ⅓ cup butter
    4 teaspoons salt
    11–13 cups flour (half Bronze Chief® whole wheat flour and
        half Natural White® flour)

Cook cereal in 2 cups water until done. Dissolve yeast in 3 cups warm water. Combine honey, butter, and salt in large bowl. Add cooked cereal to honey mixture and stir until cooled to yeast temperature. Then add dissolved yeast mixture. Add flour, stirring in 1 cup at a time. When dough is manageable, turn onto floured surface and knead 3–5 minutes, until smooth. Place in greased bowl and turn greased side up, cover with a cloth, and let rise until doubled in size.

Punch down, divide into loaves or rolls, place in pan and let rise again until doubled in size. Preheat oven to 350°.

Bake in a 350° oven until golden brown, approximately 20–25 minutes for rolls and 30–40 minutes for loaves.

Makes 4 loaves or 3 dozen rolls.

Elna Bohnen
Billings, Montana

## CINNAMON SUNRISE BREAD

2 teaspoons bread machine yeast
2 cups Prairie Gold® whole wheat flour
1⅓ cups Natural White® flour
1½ tablespoons dry milk
3 tablespoons brown sugar
1 teaspoon salt
4 tablespoons butter
2 teaspoons cinnamon
1⅓ cup water
1 cup raisins

Combine all ingredients, except raisins, in bread machine basket. Add the raisins after the rest of the ingredients have mixed for a while. Follow instructions for baking in your bread machine. I set my machine on whole wheat and select the 5-hour setting.

Makes 1 loaf.

STUART A. YATSKO
STOCKETT, MONTANA

## 3 GENERATION FINN & SLAV BREAD

2 packages active dry yeast (2 tablespoons)
1 cup warm water
2 cups milk
2 cups water (or 3 cups milk, 1 cup water, or just water)
2 tablespoons–¼ cup honey
2–5 teaspoons salt or to taste
1 tablespoon–¼ cup molasses
1 tablespoon unsweetened cocoa (optional)
1 cup Prairie Gold® or Bronze Chief® whole wheat flour (or
    more, depending how "wheaty" you want it)
½ cup Finn fiilia (yogurt) (optional)
8 cups Natural White® flour, more as needed when kneading
0–4 eggs
4–8 tablespoons butter, softened

Add yeast to warm water. Make sure it proofs out (works), 10–15 minutes (a bit of sugar can be added in to help get it going). In a saucepan, mix and warm milk, water, honey, salt, molasses, and cocoa until blended. Add 1 cup of flour to mixture (use whole wheat for this step), and let rest 15 minutes.

Add dissolved yeast (make sure mixture has cooled enough not to kill yeast), and let rest another 15 minutes.

Add yogurt (optional). Add 6 cups of flour and then add eggs (optional) and softened butter. Add and work in remaining flour (use a spatula as long as possible, then work flour in with hands while it's still in the bowl, then take the dough out and finish kneading on the counter). Knead until smooth and elastic. Set aside in greased, big bread bowl. Cover with cloth. Let rise 1–8 hours. If you like, punch it down and let rise again another hour or more.

Form loaves on greased baking sheets or in regular bread pans. Poke rows in tops of loaves with a fork. Let rise till doubled in size, 1 hour or more; can also be put in freezer and baked later.

Bake in a preheated 350° oven for 30–50 minutes, depending on size of loaf. Bread will sound a bit hollow when tapped.

Makes 2 standard loaves.

Butter a hot piece of bread; with a cold glass of milk, it can't be beat!

* * *

### HISTORY OF 3 GENERATION FINN & SLAV BREAD

Grandmother Aili Kosola Takala
∨
Mother Ramona Takala Yatsko and
Father Raymond Andrew Yatsko
∨
Stuart Andrew Yatsko

First, plant the wheat in spring. After harvest, milk the cow. Meanwhile, go out in the field and lasso a couple bees and let them lead you to honey. Supervise your maid/butler/chef as he or she mixes the ingredients. Bake, butter, enjoy!

During my summer stays on the ranch, my Grama Aili showed me how to make bread. Under her supervision, this Finn bread won a blue ribbon at the Great Falls State Fair. My parents have evolved a team cooking style where my Dad makes the bread and my Mom pans and bakes it.

Bread is very flexible and forgiving. Grama didn't use eggs or milk, though she lived on a farm. Dad likes to experiment, so he added the cocoa, yogurt, and extra salt. I like lots of molasses and honey and eggs. Experiment and enjoy!

THERESA LODE
HELENA, MONTANA

## SUNFLOWER SEED HONEY WHOLE WHEAT BREAD

1 cup *plus* 2 tablespoons warm water
3 tablespoons honey
2 tablespoons butter
1½ cups Natural White® flour
1½ cups Prairie Gold® or Bronze Chief® whole wheat flour
¼ cup sunflower seeds
1 teaspoon salt
2 teaspoons active dry yeast

Follow your breadmaker instructions to combine and bake.

Makes 1 loaf.

Frances Fenton
Sheridan, Montana

## HONEY WHOLE WHEAT BREAD

3½–4 cups Natural White® flour
2½ cups Prairie Gold® or Bronze Chief® whole wheat flour
2 packages active dry yeast
1 tablespoon salt
1 cup milk
1 cup water
½ cup honey
3 tablespoons shortening
1 egg

In mixing bowl, combine 1 cup white flour, whole wheat flour, yeast, and salt. In a saucepan, heat milk, water, honey, and shortening until warm. Add warm milk mixture to flour mixture. Add egg and, using a mixer, blend on low until moistened. Beat 3 minutes on medium speed. While running mixer on low speed, add in enough remaining all-purpose flour to make firm dough. Knead until smooth and elastic (about 5 minutes). Place in greased bowl, turning to grease top. Cover and let rise in warm place until doubled in size (about 1 hour).

Punch down dough and divide into 2 parts. Form into loaves and place in greased 9x5-inch loaf pans. Cover and let rise in warm place until doubled in size, about 30 minutes. Preheat oven to 375°.

Bake in a 375° oven for 35–40 minutes until golden brown. Remove from pans, let cool on racks.

Makes 2 loaves.

HEATHER WEIMER
FAIRVIEW, MONTANA

## WHOLE WHEAT BREAD

3 cups milk
3 tablespoons shortening
3 teaspoons salt
¼ cup honey
¼ cup molasses
¾ cup rye flour
3 cups Bronze Chief® whole wheat flour
1 package active yeast
½ cup warm water
1½ cups Prairie Gold® whole wheat flour

Heat milk, shortening, salt, honey, and molasses. Cool to lukewarm. In a large bowl, combine rye flour and 3 cups wheat flour. Dissolve yeast in warm water and add to flour mixture. Add cooled liquid. Gradually add remaining 1½ cups flour to mixture, stirring until dough pulls away from the sides of the bowl, adding more flour if needed. Knead dough with a wooden spoon for about 15 minutes or until dough is smooth and holds an imprint when pressed with finger. Cover and let rise until doubled in size, approximately 1 hour.

Knead again and let rise until doubled in size, about 30 minutes.

Divide dough into 3 loaves and place in greased pans to let rise until doubled in size. Preheat oven to 350°.

Bake in a 350° oven for 40 minutes. Remove from pan and cool.

Makes 3 loaves.

ELLABETH DEITLE
MILES CITY, MONTANA

## MUM'S BREAD

2 tablespoons shortening, melted
4 tablespoons sugar
1 tablespoon salt
3½ cups warm water
2 tablespoons active dry yeast
Natural White® flour

Combine shortening, sugar, salt, warm water, and yeast. Add white flour to desired consistency. Let rise 1 hour, punch down, and let rise for another hour. Place dough in greased pans and let rise again for 1 hour.

Bake in a preheated 350° oven for 35-45 minutes until golden.

Makes 2–3 loaves.

BOB AUERBACH
COLUMBUS, MONTANA

## WHOLE WHEAT BREAD

    1 tablespoon active dry yeast
    3 tablespoons warm water
    2¼ cups Bronze Chief® whole wheat flour
    1½ cups Natural White® flour
    2 teaspoons salt
    3 tablespoons maple syrup
    2 tablespoons walnut oil
    1⅓ cups skim milk (scalded, then cooled)

Combine yeast with warm water and proof 10 minutes. Mix together flours, salt, syrup, walnut oil, and milk, and add to proofed yeast. Knead, let rise, punch down, shape in bread pan, and let rise again. Rising times for both periods vary and should reflect prior baking experience. Grease bread pan with a little walnut oil if the pan is older and losing its non-stick properties.

Bake in a preheated 375° oven for about 35 minutes.

Makes 1 loaf.

CHERYL WILDER
FOREST PARK, GEORGIA

## BREAD MACHINE FAT-FREE FRENCH BREAD

1 cup plus 4 tablespoons water
1 teaspoon lemon juice
1 teaspoon salt
3½ cups Natural White® flour
1 package active dry yeast

Combine all ingredients in bread machine basket. Use French bread setting and follow bread machine directions for baking free-form loaves.

Makes 1 (2-pound) loaf.

LORI HENDERSON
HAVRE, MONTANA

## OATMEAL WHEAT BREAD

>  2 cups boiling water
>  1 cup rolled oats
>  2 teaspoons salt
>  1/2 cup molasses
>  2 tablespoons margarine
>  2 packages active dry yeast
>  1/3 cup warm water (105–115°)
>  2 1/2–3 cups Natural White® flour
>  1 1/2–2 cups Prairie Gold® or Bronze Chief® whole wheat flour

Pour boiling water over rolled oats. Let stand until softened. Add salt, molasses, and margarine; let cool. Soften yeast in warm water, add to oat mixture, and blend well. Gradually add flours; knead until the dough is smooth and elastic, about 10 minutes. Put into a lightly oiled bowl, turning to coat the dough on all sides. Cover; let rise for 1 hour or until doubled in size.

Punch down dough and shape into 2 loaves. Place into 2 (9x5x3-inch) loaf pans. Cover and let rise until doubled in size.

Bake in a preheated 350° oven for 35–40 minutes. Remove from pan and let cool.

Makes 2 loaves.

This is a low-fat, high-fiber bread.

Velda Hensley Welch
Toston, Montana

## DELICIOUS OATMEAL BREAD

1½ cups boiling water
1 cup rolled oats
¾ cup molasses
3 tablespoons butter, softened
2 teaspoons salt
1 tablespoon active dry yeast
2 cups warm water
8 cups Prairie Gold® whole wheat flour

Pour boiling water over the oats and let stand 30 minutes.

Add molasses, butter, and salt to oat mixture. Dissolve yeast in warm water and add to oat mixture. Beat and work in enough of the flour to make a medium soft dough. Turn onto floured board and knead until smooth, about 10 minutes. Place the dough in a greased bowl, cover, and let rise in a warm place until doubled in size, about 1 hour.

Turn dough onto board and knead again. Divide and shape into 2 loaves. Use 2 greased (9x5-inch) loaf pans. Cover and let rise until doubled in size, about 45 minutes.

Preheat oven to 400°. Bake for 5 minutes, then lower heat to 350° and bake for 40 minutes or until loaves sound hollow when tapped. Remove bread from the oven and brush tops with butter and eat while warm!

Makes 2 loaves.

DEENA GOBBS
TOWNSEND, MONTANA

## CRACKED WHEAT BREAD

2 packages active dry yeast
1/2 cup warm water
2 cups 100% cracked wheat
5 cups Natural White® flour
1/2 cup brown sugar
2 tablespoons butter
1 tablespoon salt
2 cups boiling water

Soften yeast in 1/2 cup warm water for 10 minutes. Combine cracked wheat, 2 cups flour, brown sugar, butter, and salt, and pour boiling water over mixture. Cool to warm. Add softened yeast to cracked wheat mixture and gradually beat in rest of flour to make stiff dough. Knead for 8–10 minutes. Place in a greased bowl. Let rise 1 1/2 hours in warm place.

Punch down dough, divide into 2 parts, and knead into loaves. Place dough into 2 greased pans. Let rise until doubled in size, about 1 1/2 hours.

Bake in a preheated 400° oven for 30 minutes.

Makes 2 loaves.

Betty Horne
Townsend, Montana

## SHORT-CUT YEAST DOUGH MIX

> 1 (5-pound) bag Natural White® flour or Prairie Gold®
> whole wheat flour
> 2½ tablespoons salt
> 1 cup sugar
> 2 cups nonfat dry milk

In large bowl, combine and stir all ingredients until well blended. Seal in tightly covered container or in heavy plastic bags and store in cool place for up to 1 month, or store refrigerated for up to 6 months. Stir dough mix well before each use. When ready to bake, measure out mix and add yeast, egg, and oil for most recipes.

To use Prairie Gold® flour, use 5 pounds *plus* 4 extra cups.


## BASIC LOAF RECIPE

> 1 package active dry yeast
> 1 cup warm water (110°)
> 2 tablespoons margarine or butter, melted
> 1 egg
> 3½ cups Short-Cut Yeast Dough Mix (see above)

Dissolve yeast in water. Stir in margarine and egg to dissolved yeast. Add prepared yeast dough mix and blend well. Knead, add flour as needed to prevent sticking. Place dough in a greased bowl, turn over to coat. Let dough rise for about one hour until doubled in size, then knead lightly. Shape loaf and place in greased pan. Let dough rise in pan for about 45 minutes. Bake in a preheated 350° oven for about 30 minutes. Cool on a rack.

Makes 1 loaf.

BETTY HORNE
TOWNSEND, MONTANA

---

## THREE-WHEAT BATTER BREAD

1 package active dry yeast
½ cup warm water
⅛ teaspoon ground ginger
3 tablespoons honey
1 (13-ounce) can evaporated milk
1 teaspoon salt
2 tablespoons salad oil
2½ cups Natural White® flour, unsifted
1¼ cup Prairie Gold® or Bronze Chief® whole wheat flour, unsifted
½ cup wheat germ
¼ cup cracked wheat

In large bowl, combine yeast, water, ginger, and 1 tablespoon honey; let stand in warm place until bubbly, about 20 minutes. Stir in remaining honey, milk, salt, and oil. In a separate bowl, stir together flours, wheat germ, and cracked wheat. Add flour mixture to liquid ingredients, 1 cup at a time, beating after each addition until well blended. Spoon batter evenly into a well-greased, 2-pound coffee can or into 2 well-greased, 1-pound coffee cans; cover with greased plastic lids. (Freeze if you wish to use later.) Let rise in a warm place until lids pop off (about 55–60 minutes for 1-pound cans, about 1–1½ hours for 2-pound cans). (If batter is frozen, let stand in cans at room temperature until lids pop off—4–5 hours for 1-pound can and 6–8 hours for 2-pound cans.)

Bake in cans without lids in a preheated 350° oven for 45 minutes for 1-pound cans, about 60 minutes for a 2-pound can, or until bread sounds hollow when tapped. Let cool in cans on racks for 10 minutes; then loosen crust around edge of can with a thin knife, slide bread from can, and let cool in an upright position on rack.

Makes 1 large or 2 small loaves.

DEENA GOBBS
TOWNSEND, MONTANA

## YEAST WHOLE WHEAT MUFFINS

1 package active dry yeast
1 cup warm water
½ cup honey
½ cup cooking oil
½ teaspoon salt
3 cups Prairie Gold® whole wheat flour
1 egg, beaten
Sesame seeds

Dissolve yeast in warm water and let sit while creaming together honey, oil, and salt. Add 1 cup flour to creamed mixture. Add yeast mixture, egg, and remaining flour. Mix well and keep refrigerated until needed. Put sesame seeds in bottom of muffin pan before putting in batter.

Preheat oven to 375°. Let muffins rise until doubled in size and bake in a 375° oven for 15 minutes.

Makes 15–18 muffins.

SHIRLEY MARKS
GREAT FALLS, MONTANA

## NEW YORK WATER BAGELS

$1\frac{1}{8}$ cups water
2 teaspoons active dry yeast
1 tablespoon vegetable oil
2 tablespoons malt syrup, molasses, or white sugar
1 teaspoon salt
$3\frac{1}{3}$ cups Natural White® flour

Combine water and yeast. Then add and mix together remaining ingredients (using only 3 cups of the flour) and knead at least 10 minutes, adding final $\frac{1}{3}$ cup flour when dough becomes sticky. In a lightly greased, covered container, let the dough rise until doubled in size. This usually takes about an hour. Test by poking two fingers lightly and quickly about $\frac{1}{2}$ inch into the dough. If the dent stays, the dough is doubled. Spray baking sheets with a non-stick vegetable spray or spread with a little vegetable oil.

To form dough into bagels, separate into 8–10 pieces. Roll each piece into a ball and shape into a bagel by making a hole with 2 floured fingers. Twirl the circle around your index fingers, like a hula hoop. Pull out and shape each round. Place bagels onto greased baking sheet and cover with a sheet of greased plastic or a dampened tea towel. Allow them to rise at room temperature until puffy, about 20 minutes.

Preheat oven to 400°. Fill a 4–6-quart pot with water 3–4 inches deep. Drop bagels one at a time into boiling water. Boil about four at a time. Boil for about 30 seconds on each side turning with a slotted spatula. Place back on baking sheet. At this time, toppings can be added, such as seeds, nuts, etc.*

Bake in a 400° oven for 20–25 minutes or until tops are golden brown.

*To add toppings, beat 1 egg white with $1\frac{1}{2}$ teaspoons water. Brush on bagel tops with a pastry brush and add poppy seeds, sesame seeds, etc.

## SHIRLEE GATES
## COLUMBUS, MONTANA

---

## MUM'S BUNS

1½ cups lukewarm water
¼ cup sugar
1 package active dry yeast
2 eggs, well beaten
1 teaspoon salt
¼ cup vegetable oil
1 teaspoon vinegar
5 cups Natural White® flour

In a large bowl, combine the water, sugar, and yeast. When the yeast rises to the top of the mixture, add eggs, salt, oil, and vinegar. Mix in flour; knead well. Let rise only once until doubled in size, about 1–1½ hours.

Preheat oven to 375°. Form dough into small buns and arrange on a greased baking pan; let rise, about ½ hour or until doubled. Bake 18–20 minutes in a 375° oven. Remove from oven and brush with butter.

Makes about 28 buns.

FRANCES FENTON
SHERIDAN, MONTANA

## WHOLE WHEAT SANDWICH BUNS

2 cups water
1/2 cup white sugar
1/2 cup nonfat dry milk
1 tablespoon salt
3/4 cup cooking oil
41/2–5 cups Natural White® flour
31/2 cups Bronze Chief® whole wheat flour
2 packages active dry yeast
3 eggs, beaten
Milk

Combine water, sugar, dry milk, salt, and oil in saucepan. Heat to warm (120–130°). Stir together 4 cups white flour, 31/2 cups wheat flour, and yeast in a mixer bowl. Add the warm liquids and beaten eggs. Beat at low speed for 1–2 minutes, scraping sides of bowl. Beat on high speed 3 minutes, scraping bowl occasionally. Add only enough remaining white flour to make moderately soft dough. Knead until smooth and elastic, about 5 minutes. Place in greased bowl; turn to grease top. Cover and let rise until doubled in size, about 11/2 hours.

Punch down. Divide dough into thirds. Cover and let rest 5 minutes. Divide each third into 8 portions. Shape into balls and place on greased baking sheet, press down with palms of hands into 31/2–4-inch rounds. Cover and let rise until doubled in size (30–40 minutes).

Preheat oven to 375°. Brush rounds with milk. Bake in a 375° oven for 12 minutes or until golden brown. Brush with butter.

Makes 24 buns.

PAM HUCKINS
BOISE, IDAHO

## CINNAMON ROLLS

### Dough:
2 packages active dry yeast
2 cups very warm water (110-115°)
½ cup sugar
2 teaspoons salt
6½–7 cups Natural White® Flour
1 egg
¼ cup butter-flavored shortening

### Filling:
½ cup (1 stick) butter, very soft
½ cup brown sugar
½ cup granulated sugar
3 tablespoons cinnamon

In a mixing bowl dissolve yeast in water. Add sugar, salt, and half of the flour; beat for 2 minutes. Add egg and shortening. Gradually beat in rest of the flour until smooth, if using dough hook attachments. If not, mix in as much flour as possible by machine; continue to add remaining flour by stirring in with wooden spoon. Place dough in lightly greased bowl. Cover with damp cloth and refrigerate for 8 hours or overnight.

Two hours before baking time: Remove dough from refrigerator, punch down. Let rise 30–40 minutes.

Preheat oven to 400°. Roll dough out on floured surface to a 15x20-inch rectangle. Spread softened butter over dough. Sprinkle sugars, followed by cinnamon over entire surface. Roll dough in jellyroll style. Pinch and seal end seams. Cut with bread knife into 2-inch-thick rolls. Place cut rolls in a greased 9x13-inch pan, about ½ inch apart. Bake rolls in a 400° oven for 15–20 minutes. Frost if desired.

Makes 12–16 rolls.

CONNIE FRETHEIM
SHELBY, MONTANA

## ICE BOX CINNAMON ROLLS

### Dough:
2 packages active dry yeast
1/2 cup warm water
2 cups lukewarm milk (scalded then cooled)
1/3 cup sugar
1/3 cup shortening or vegetable oil
3 teaspoons baking powder
2 teaspoons salt
1 egg
5–6 cups Prairie Gold® whole wheat flour

### Filling:
1/2 cup sugar
1 tablespoon cinnamon
4 tablespoons butter, softened

### Frosting:
1 cup powdered sugar
1 tablespoon milk
1/2 teaspoon vanilla

Dissolve yeast in warm water. Stir in milk, 1/3 cup sugar, shortening, baking powder, salt, egg, and 2–3 cups of the flour. Beat until smooth. Mix in remaining flour to make dough easy to handle. Turn dough onto a well-floured surface, knead until smooth and elastic (8–10 minutes). Place in greased bowl; turn the dough once in the bowl to coat. Cover and let rise in warm place until doubled in size (1½ hours).

Grease 2 (9x13-inch) pans. Punch down dough, divide into halves. Roll half the dough into a 10x12-inch rectangle. Mix 1/2 cup sugar with cinnamon. Spread rectangle with 2 tablespoons butter and sprinkle with half the sugar-cinnamon mixture. Roll up dough beginning at wide side. Pinch edge of dough into roll to seal. Cut roll into 12 slices. Place rolls slightly apart in one pan. Wrap pan tightly with heavy foil. Repeat with remaining dough. Refrigerate at least 12 hours but no longer than 48 hours.

Bake in a preheated 350° oven for 30–35 minutes. Combine frosting ingredients well. Remove rolls from oven and frost with powdered sugar frosting.

Makes 24 rolls.

**VARIATION:**

**Caramel Pecan Icebox Rolls**

Omit powdered sugar frosting. Before rolling dough into rectangle, heat 1 cup brown sugar and $\frac{1}{2}$ cup light corn syrup until it resembles caramel. Divide caramel mixture between the two pans. Sprinkle each with $\frac{1}{2}$ cup pecan halves. Roll dough, slice, refrigerate, and bake as directed above. Invert pan immediately on a large tray; serve.

# Desserts

*Raspberry Cream Cheese Coffee Cake*
*(see page 108)*

WHEAT MONTANA
THREE FORKS, MONTANA

## IDENTITY PRESERVED WHEAT COOKIES

### Dough:
1 cup shortening
2 eggs
½ cup brown sugar
1¼ teaspoons vanilla
2 cups Natural White® flour
½ teaspoon salt
2 egg whites, slightly beaten
1½ cups finely chopped nuts

### Thumbprint filling:
Any flavor jam
½ cup Cooked Wheat (see page 144)

Preheat oven to 350°.

Combine and mix shortening, eggs, brown sugar, and vanilla. Sift together flour and salt, and add to shortening mixture to form dough. Roll dough into 1-inch balls, dip into egg whites, and roll to coat with chopped nuts. Place coated dough balls 1 inch apart on an ungreased cookie sheet. Bake in a 350° oven for 5 minutes. Remove from oven and quickly and gently press thumb on top of cookie. Return to heated oven for 8 minutes. Cool.

Combine jam and Cooked Wheat to make filling. Spoon filling into thumbprint of each cooled cookie.

Makes approximately 12 cookies.

WHEAT MONTANA
THREE FORKS, MONTANA

## PEANUT COOKIES

½ cup butter or margarine, softened
½ cup peanut butter
½ cup sugar
½ cup brown sugar
1 egg
1½ cups Prairie Gold® whole wheat flour
½ teaspoon baking powder
¾ teaspoon baking soda
¼ teaspoon salt
Flour (for fork dipping)

Preheat oven to 375°.

Beat butter, peanut butter, sugars, and egg together until light and fluffy. Stir flour, baking powder, baking soda, and salt into peanut butter mixture. Shape batter into walnut-sized balls and place onto a lightly greased baking sheet. Dip a fork in flour and flatten dough ball with tines one way, then the other. Bake in a 375° oven until golden brown.

Makes approximately 12 cookies.

SHERYL KNOWLES
GREAT FALLS, MONTANA

## CARROT–APPLE CAKE

4 eggs
1 cup sugar
½ cup oil
1 cup Prairie Gold® whole wheat flour
1 cup Natural White® flour
2 teaspoons baking soda
1 teaspoon salt
3 teaspoons cinnamon
1½ cups grated carrots
1½ cups grated apples
1 cup nuts or raisins (either or none as desired)

In mixing bowl, beat eggs, sugar, and oil. Add dry ingredients and mix. Stir in carrots, apples, and nuts or raisins. Pour ingredients into a 9x13-inch greased pan or a rimmed cookie sheet for snack cake. Bake in a preheated 325° oven for 40–50 minutes. When cooled, frost with Cream Cheese Frosting.

Makes 15–24 servings, depending on pan used.

## CREAM CHEESE FROSTING

3 ounces cream cheese
4 tablespoons butter
3 cups powdered sugar
½ teaspoon vanilla

Soften cream cheese and butter. Add powdered sugar and vanilla, and beat until fluffy.

Makes enough frosting for 9x13-inch cake.

PENNY CLIFTON
HUNTLEY, MONTANA

## RUSK CAKE

16 cups of Natural White® flour
3 tablespoons baking powder
1 teaspoon salt
4½ cups (9 sticks) butter/margarine
1¾ cups brown sugar
6 cups bran flakes cereal
4 cups bran
1 cup sunflower seed meats
1 cup raisins
4 eggs, beaten
3 cups buttermilk

Preheat oven to 450°.

Sift together flour, baking powder, and salt. Cut in butter. Add brown sugar. Mix in bran flakes cereal, bran, sunflower seed meats, raisins, eggs, and buttermilk. Pour batter into a greased, large, deep, rectangular cake pan (as for sheet cakes) and bake for 1 hour in a 450° oven. Cool. Cut into small squares and remove from pan. Let cake pieces sit out until very dry.

Note: I brought this recipe back from South Africa, where I had been visiting friends. A "rusk" is a hardened cake that keeps forever and is usually dipped in a hot beverage. Great for camping. Use a BIG bowl and your hands for this hefty recipe!

Makes about 32 (2x2-inch) pieces.

Dannon Giles
Missoula, Montana

## GINGER MOLASSES COOKIES

½ cup brown sugar
½ cup shortening
½ cup molasses
3½ cups Natural White® flour
1 teaspoon baking soda
½ teaspoon salt
1 teaspoon ground ginger
2 teaspoons cinnamon
½ cup sour milk
½ teaspoon vinegar

Gradually cream sugar into shortening. Stir in molasses. Sift together 1¼ cups flour, baking soda, salt, ginger, and cinnamon, and add to molasses mixture. Stir thoroughly. Combine milk and vinegar. Add remaining flour alternately with milk-vinegar mixture to rest of dough. Chill dough for at least 2 hours.

Preheat oven to 375°. Roll out dough to about ¼-inch-thick on floured board. Cut into shapes using cookie cutters. Place on baking sheet (greased and floured) and bake in a 375° oven for 10–12 minutes.

Makes about 36 cookies.

ALANE FITZPATRICK
HELENA, MONTANA

## STRAWBERRIES 'N CREAM BLONDIES

### Cream cheese mixture:
4 ounces light cream cheese, softened
1/2 teaspoon grated lime rind
1 tablespoon lime juice
1/2 cup reduced-sugar strawberry preserves or strawberry jam
1 egg

### Brownie mixture:
2 (1-ounce) squares white baking chocolate
1/2 cup (1 stick) light margarine or butter
1 cup packed brown sugar
1/2 cup granulated sugar
2 eggs
1 teaspoon vanilla
1 1/4 cups Natural White® flour

Preheat oven to 350°.

With an electric mixer, beat cream cheese, lime rind, and lime juice. Add strawberry jam and beat thoroughly. Add egg and mix until smooth. Refrigerate.

Melt white chocolate squares and margarine in a medium saucepan over low heat. Using a wooden spoon, add brown sugar and granulated sugar, stirring well. Mix in eggs and vanilla. Stir in flour until mixture is smooth.

Pour brownie mixture into a 13x11x2-inch baking dish that has been coated with non-stick cooking spray. Remove cream cheese mixture from refrigerator and drop by tablespoons onto brownie mixture in pan. With a knife, swirl cream cheese mixture into brownie mixture.

Bake for 30 minutes or until done (test center with wooden toothpick). Cool and cut into bars.

Makes 24 bars.

SHIRLEE GATES
COLUMBUS, MONTANA

## PRAIRIE GOLD® CRUMB CAKE

¾ cup (1½ sticks) butter
1 cup dark brown sugar
2 cups Prairie Gold® whole wheat flour
1 teaspoon baking soda
¼ teaspoon cloves
¼ teaspoon nutmeg
½ teaspoon cinnamon
1 egg
1 cup sour milk or buttermilk

Preheat oven to 350°.

Cream butter and sugar, and beat until fluffy. Add flour, cutting it in with a pastry blender until mixture forms very fine crumbs. Reserve 1 cup of this mixture for the topping. To the remainder add soda and spices. Beat egg and milk until blended. Add to spiced mixture and mix until well blended. Spread batter into a well-greased 8x8-inch pan and sprinkle reserved crumbs on top. Spread evenly and pat gently into batter. Bake in a 350° oven for 40–50 minutes.

Serves 6.

BETTY KELLER
FAIRFIELD, IOWA

## OLD-FASHIONED SUGAR COOKIES

1 cup (2 sticks) oleo
1¾ cups sugar
1 teaspoon vanilla
3 eggs
¼ teaspoon soda
8 ounces sour cream
4 teaspoons baking powder
5–5½ cups Natural White® flour

Preheat oven to 350°.

Thoroughly cream oleo and sugar. Add vanilla. Add eggs and beat until light and fluffy. Mix soda and sour cream together and add to creamed mixture. Sift together baking powder and flour and blend into creamed mixture. Dough should be easy to handle and roll out.

On a lightly floured surface, roll dough to ⅛-inch thickness. Cut into desired shapes with cutters. Sprinkle cookies with sugar. Bake on an ungreased cookie sheet in a 350° oven for 9 minutes. These are also good frosted.

Makes about 4 dozen.

Peggy Howe
Billings, Montana

## CHOCOLATE CHIP PEANUT BUTTER COOKIES

½ cup (1 stick) butter or margarine
½ cup soft shortening
1 cup crunchy peanut butter
1 cup granulated sugar
1 cup brown sugar
2 eggs
2½ cups Natural White® flour
1½ teaspoons baking soda
1 teaspoon baking powder
½ teaspoon salt
2 cups milk chocolate chips

Preheat oven to 375°.

In large bowl, beat butter, shortening, peanut butter, granulated sugar, brown sugar, and eggs on medium speed until well blended. Stir together flour, baking soda, baking powder, and salt. Gradually add sifted ingredients to butter mixture, beating until well blended. Stir in chocolate chips. Drop by rounded teaspoons onto ungreased cookie sheet. Bake 8–10 minutes in a 375° oven or until set. Cool slightly. Remove from cookie sheet.

Makes about 6 dozen cookies.

THERESA LODE
HELENA, MONTANA

## APPLE CRISP

### Filling:
About 10 apples, peeled and sliced
¾ cup brown sugar
1 teaspoon cinnamon
¼ teaspoon nutmeg
¼ teaspoon salt
¼ teaspoon ground ginger
¼ teaspoon ground coriander
¼ cup Natural White® flour

### Topping:
½ cup Prairie Gold® whole wheat flour
½ cup quick oats
½ cup brown sugar
½ teaspoon cinnamon
Pinch of salt
½ cup (1 stick) butter, melted

Preheat oven to 350°.

Mix all apple filling ingredients and place in a 9x9-inch greased baking dish.

Combine all topping ingredients. Using a light touch so it doesn't clump up too much, sprinkle the topping over the filling. Bake in a 350° oven for about 40 minutes.

Serve with vanilla ice cream while still warm.

Note: I usually double the topping, even with a small batch of filling—your family will like the extra crunchies!

Serves 9.

LEIGH SMITH
BILLINGS, MONTANA

## MEGAN'S SCOTTISH SCONES

2 cups Natural White® flour, unsifted
½ cup sugar
2 teaspoons cream of tartar
1 teaspoon baking soda
¾ teaspoon salt
½ cup shortening
½–1 cup raisins, currants, or Craisins (optional)
2 eggs, slightly beaten
¼ cup milk

Preheat oven to 350°.

Sift dry ingredients into a large bowl. Cut in shortening thoroughly. Add remaining ingredients and mix with a fork—do not knead dough. Divide dough in half. Flatten, but don't roll, each half into a small circle, about ½-inch thick. Cut each half into 8 "triangles."

Bake on a greased and floured baking sheet in a 350° oven for 15 minutes or until done.

Makes 16 scones.

DORIS FALK
LITTLE FALLS, MINNESOTA

## SWEET CORN–APPLESAUCE CUPCAKES

1⅓ cups Natural White® flour
⅔ cup sugar
1 teaspoon baking powder
1 teaspoon cinnamon
¼ teaspoon nutmeg
½ teaspoon salt
2 eggs
1 cup raw "super sweet" sweet corn (about 3 cobs)*
⅓ cup vegetable oil
½ cup applesauce

Preheat oven to 350°.

Mix dry ingredients. Stir in remaining ingredients until just moistened. Fill paperlined muffin cups half full with batter. Bake in a 350° oven for 30 minutes. Cool. Frost with favorite frosting.

Makes 24 cupcakes.

*The "super-sweet" sweet corn is crunchy and takes the place of nuts in these cupcakes.

Margaret Caster
Worden, Montana

---

## APPLE ROLLS

>   4 cups Natural White® flour
>   2 tablespoons granulated sugar
>   2 tablespoons baking powder
>   1 teaspoon salt
>   2½ tablespoons shortening
>   1 egg
>   Milk
>   6 medium-sized apples, peeled, cored, and sliced
>   Apple Rolls Syrup, warmed (see below)

Preheat oven to 375°.

Combine dry ingredients. Cut in shortening until the mixture resembles small peas. Beat egg in cup, then add milk to make 1 cup. Mix milk-egg mixture into flour crumbs until dough forms. Roll out dough to ¼-inch thickness and cover with sliced apples. Roll up like a jellyroll and cut in 1-inch slices. Place apple rolls on greased 9x13x2-inch baking pan and pour hot Apple Rolls Syrup over. Bake in a 375° oven for 35–40 minutes.

Serves 6–8.

## APPLE ROLLS SYRUP

>   2 cups brown sugar
>   2 cups water
>   4 tablespoons butter or margarine
>   2 tablespoons Natural White® flour

Mix together all ingredients and bring to a boil, stirring until smooth. Simmer for 3 minutes. Use hot.

CLAUDIA REES
LIVINGSTON, MONTANA

## GOLDEN TREASURE PIE

2 (8½-ounce) cans crushed pineapple, undrained
½ cup *plus* ⅔ cup sugar
2 tablespoons cornstarch
2 tablespoons water
1 tablespoon butter
¼ cup Natural White® flour
1 cup cottage cheese
1 teaspoon vanilla
½ teaspoon salt
2 eggs, slightly beaten
1¼ cup milk
10-inch pie shell, unbaked

Preheat oven to 450°.

Combine pineapple, ½ cup sugar, cornstarch, and water in a small saucepan. Bring to a boil. Cook 1 minute, stirring constantly. Cool. In a mixing bowl, blend ⅔ cup sugar and butter. Add flour, cottage cheese, vanilla, and salt. Beat until smooth. Slowly add eggs then milk to cottage cheese mixture, beating constantly. Pour pineapple mixture into unbaked pie crust, spreading evenly. Gently pour custard over pineapple being careful not to disturb first layer. Bake at 450° for 15 minutes, then reduce heat to 325° and bake 45 minutes longer. Serve cold.

Makes 8–10 servings.

Note: Pineapple layer can be made in advance.

CLAUDIA REES
LIVINGSTON, MONTANA

## WET BOTTOM SHOOFLY PIE

1 cup Natural White® flour
2/3 cup brown sugar
1 tablespoon butter
1 cup dark Karo syrup or mild flavor molasses
3/4 cup boiling water
1 egg, beaten
1 teaspoon baking soda
9-inch unbaked pie shell

Preheat oven to 375°.

Combine and mix the flour, brown sugar, and butter. Reserve 1/2 cup of this crumb mixture for the topping. To the remaining crumbs, add the molasses, boiling water, egg, and baking soda; mix well. Put the molasses mixture into the unbaked pie shell. Sprinkle the 1/2 cup of reserved crumbs on top. Bake for 10 minutes in a 375° oven, then reduce heat to 350° and continue baking for 30 minutes. Cool on rack.

Serves 8.

CARLEEN EISELE
DIXON, ILLINOIS

## BANANA CAKE

2½ cups Natural White® flour
1½ cups sugar
1½ teaspoons baking powder
1 teaspoon baking soda
1 teaspoon salt
½ cup shortening, softened
1 cup mashed ripe banana
⅔ cup buttermilk
2 eggs
1 teaspoon vanilla

Preheat oven to 350°.

Combine flour, sugar, baking powder, baking soda, and salt. Add in shortening and banana. Mix in buttermilk, eggs, and vanilla. Beat batter for 2 minutes. Pour into greased, floured 9x13-inch cake pan. Bake in a 350° oven for 40 minutes.

Cool, then frost with cream cheese or chocolate frosting.

Serves 16.

CARLEEN EISELE
DIXON, ILLINOIS

## GINGERSNAPS

1 cup sugar
1 cup shortening
1 egg
1/4 cup molasses
2 1/2 teaspoons baking soda
1/4 teaspoon salt
1 1/2 cups Natural White® flour
1/2 cup Bronze Chief® whole wheat flour
1 teaspoon cinnamon
1 teaspoon cloves
1/2 teaspoon dried, ground ginger
Granulated sugar for coating

Cream together sugar and shortening. Add egg, molasses, baking soda, and salt; blend well. In a separate bowl, combine flours, cinnamon, cloves, and ginger. Mix wet and dry ingredients. Refrigerate batter for 30 minutes.

Preheat oven to 350°. Shape batter into balls using teaspoon and roll in granulated sugar. Place on ungreased cookie sheet (do not press). Bake in a 350° oven for 8–10 minutes. Cool on cookie sheet for 5 minutes, then remove to wire rack.

Makes 4 dozen cookies.

CARLEEN EISELE
DIXON, ILLINOIS

## CHOCOLATE BUTTERMILK CAKE

2 cups sugar
2 cups Natural White® flour
1 teaspoon baking soda
1 cup (2 sticks) butter
1 cup water
4 teaspoons cocoa
½ cup buttermilk
2 eggs
1 teaspoon vanilla

Preheat oven to 350°.

Combine sugar, flour, and baking soda in a mixing bowl. In a small saucepan, combine butter, water, and cocoa. Bring chocolate mixture to a boil, then cool slightly. When cooled, blend chocolate mixture into dry ingredients. Add buttermilk, eggs, and vanilla, and beat well. Pour batter into a greased, floured 11x17-inch jellyroll pan. Bake in a 350° oven for 20 minutes. Cool and then frost (see below).

Serves 16.

## FROSTING

½ cup cocoa
3 cups powdered sugar
6 tablespoons butter
4–5 tablespoons milk
1 teaspoon vanilla

Combine all ingredients well.

Deon Tomsheck
Ethridge, Montana

## WHEAT MONTANA POPPY SEED CAKE

$1\frac{1}{2}$ cups Natural White® flour
$1\frac{1}{2}$ cups Prairie Gold® whole wheat flour
$\frac{1}{3}$ cup poppy seeds
$2\frac{1}{2}$ teaspoons baking soda
$\frac{1}{2}$ teaspoon salt
$\frac{3}{4}$ cup ($1\frac{1}{2}$ sticks) butter or margarine
$1\frac{1}{2}$ cups honey
1 teaspoon vanilla
4 eggs
$\frac{1}{2}$ cup buttermilk or sour milk
1 small banana, mashed
$\frac{1}{2}$ cup raisins (optional)

Preheat oven to 350°.

Grease and lightly flour a 10-inch fluted tube pan or a 9x13-inch pan. Combine the first 5 ingredients. Beat butter about 30 seconds. Add honey and vanilla to butter; beat until fluffy. Add eggs, one at a time, beating 1 minute after each. Combine buttermilk and banana. Add dry ingredients and buttermilk mixture alternately to honey mixture, beating after each addition. Stir in raisins. Pour batter into pan and spread evenly. Bake in a 350° oven for 45–55 minutes or until done. Cool 15 minutes in pan on wire rack. Invert onto a wire rack and remove pan. Cool thoroughly.

Serves 12.

Deon Tomsheck
Ethridge, Montana

## 7-GRAIN CAKE

1 cup Wheat Montana 7-grain cereal
1½ cups boiling water
½ cup (1 stick) butter, softened
1 cup sugar
1 cup packed brown sugar
2 eggs
1 teaspoon vanilla extract
1½ cups Natural White® flour
1½ teaspoons baking soda
1 teaspoon cinnamon
1 teaspoon nutmeg
½ teaspoon salt

Preheat oven to 350°.

In a large bowl, combine 7-grain cereal and boiling water. Let stand for 20 minutes. Cream butter and sugars in bowl until light and fluffy. Add eggs and vanilla; mix well. Sift together flour, baking soda, cinnamon, nutmeg, and salt, and add to creamed mixture. Mix in prepared 7-Grain cereal. Pour batter into a greased, floured 9-inch square pan. Bake in a 350° oven for 45–50 minutes or until cake tests done. Cool and cut into squares.

Serves 9.

DEON TOMSHECK
ETHRIDGE, MONTANA

## DEON'S $200 COOKIES

2½ cups Wheat Montana 7-grain cereal
1 cup (2 sticks) butter
1 cup sugar
1 cup brown sugar
2 eggs
1 teaspoon vanilla
2 cups Natural White® flour
½ teaspoon salt
1 teaspoon baking powder
1 teaspoon baking soda
6 ounces chocolate chips
6 ounces butterscotch chips
1 (4-ounce) sweet chocolate bar, grated
1½ cups chopped nuts

Preheat oven to 375°.

Process 7-grain cereal in blender to a fine powder. Cream butter and both sugars
together. Add eggs and vanilla. Mix together with flour, 7-grain cereal powder,
salt, baking powder, and baking soda. Add chips, chocolate bar, and nuts. Roll
into balls and place 2 inches apart on a baking sheet. Bake in a 375° oven for 6
minutes.

Makes 5–6 dozen cookies.

Pam Huckins
Boise, Idaho

## EASY BLACK FOREST CAKE

6 squares semisweet baking chocolate
¾ cup (1½ sticks) butter
1½ cups sugar
3 eggs
2 teaspoons vanilla
2½ cups Natural White® flour
1 teaspoon baking soda
¼ teaspoon salt
1½ cups water
1 pint heavy whipping cream
½ teaspoon vanilla
2 tablespoons sugar
1 (29-ounce) can of cherry pie filling

Preheat oven to 350°.

Melt chocolate and butter together in large microwavable bowl for 2 minutes, or until butter is melted. Stir until chocolate is completely melted. Stir sugar into melted chocolate mixture until well blended. Beat in eggs, one at a time, with electric mixer until mixed. Add vanilla, ½ cup flour, baking soda, and salt; mix well. Beat in the remaining 2 cups of flour alternately with water until smooth. Pour batter into 2 greased, floured 9-inch layer pans. Bake in a 350° oven for 35 minutes or until  toothpick inserted into center comes out clean.

Cool in pans for 10 minutes. Remove from pans to cool on wire racks. When cake is completely cooled, whip whipping cream with vanilla and sugar. Spread top of bottom cake layer with ½–¾ of the can of cherry pie filling. Place top cake over cherry pie filling. Frost the rest of cake with whipping cream. Gently spoon remaining cherry pie filling on top of cake for garnish.

Serves 12.

MARILYN STEINGRUBER
MANHATTAN, MONTANA

## RASPBERRY CREAM CHEESE COFFEE CAKE

2¼ cups Natural White® flour
¾ cup sugar
¾ cup (1½ sticks) butter
½ teaspoon baking powder
½ teaspoon baking soda
½ teaspoon salt
¾ cup sour cream
1 egg, beaten
1½ teaspoons almond extract

**Filling:**
1 (8-ounce) package cream cheese, softened
½ cup sugar
1 egg

**Topping:**
½ cup raspberry jam
½ cup slivered almonds

Preheat oven to 350°.

Combine flour and sugar. Cut in butter. Remove 1 cup of flour mixture and set aside. To the remaining flour mixture add baking powder, baking soda, salt, sour cream, egg, and almond extract. Mix well. Spread batter into bottom and up sides of 9-inch springform pan.

For the filling, beat cream cheese, sugar, and egg in a small bowl. Mix well and pour over batter. Spoon jam on top of filling. Sprinkle jam layer with almonds and reserved crumbs. Bake in a 350° oven for 1 hour. Let stand 15 minutes before removing sides from pan.

Serves 12.

MARILYN STEINGRUBER
MANHATTAN, MONTANA

## MOUNTAIN COOKIES

1 cup (2 sticks) butter
1 cup powdered sugar
2 teaspoons vanilla
2 cups Natural White® flour
1/2 teaspoon salt

### Filling:

1 (3-ounce) package cream cheese, softened
1 cup powdered sugar
2 tablespoons Natural White® flour
1 teaspoon vanilla
1/2 cup pecans (optional)

### Topping:

1/2 cup chocolate chips
2 tablespoons butter
2 tablespoons water
1/2 cup powdered sugar

Preheat oven to 350°.

In a mixing bowl, combine and cream the butter, sugar, and vanilla. Combine flour and salt; add to creamed mixture and mix well. Roll dough into balls and place on ungreased cookie sheets. Make a deep indentation in each cookie. Bake in a 350° oven for 10–12 minutes or until edges just start to brown, and cool completely.

For the filling, beat cream cheese, sugar, flour, and vanilla in a mixing bowl. Add pecans and mix well. Spoon prepared filling into each cookie.

For the topping, heat chocolate chips, butter, and water in a small saucepan until melted; stir in sugar. Drizzle over cookies. Cool cookies until chocolate topping is set.

Makes about 24 cookies.

Marilyn Steingruber
Manhattan, Montana

## TWEEDIES

½ cup (1 stick) butter
⅔ cup sugar
1⅓ cups Natural White® flour
2 teaspoons baking powder
½ teaspoon salt
1 cup milk
2 squares semisweet chocolate, grated
1 teaspoon vanilla
2 egg whites, beaten stiff

**Topping:**
⅓ cup butter
2 egg yolks
2 cups powdered sugar
1 teaspoon vanilla

**Glaze:**
1 tablespoon oil
2 squares semisweet chocolate
Walnuts, crushed

Preheat oven to 350°.

Cream butter and sugar. Combine flour, baking powder, and salt. Add dry ingredients alternately with milk and grated chocolate to the creamed mixture. Add vanilla and egg whites. Pour batter on a baking sheet with sides. Bake in a 350° oven for 30 minutes. Cool.

Mix all topping ingredients and spread on top of cookie bars.

Melt oil with chocolate for glaze. Drizzle glaze for ribbon or marble effect over icing on cake. Sprinkle with crushed walnuts.

Makes 1 pan of Tweedies.

CAROL GARLOW
LIVINGSTON, MONTANA

## CHOCOLATE PEANUT BUTTER BROWNIES

1 cup *plus* 1 tablespoon Natural White® flour
1/4 teaspoon baking powder
1/2 teaspoon salt
3/4 cup (1 1/2 sticks) unsalted butter
3 ounces unsweetened chocolate, finely chopped
3 eggs
1 1/3 cups packed dark brown sugar
1 1/2 teaspoons pure vanilla
3/4 cup creamy peanut butter
1/4 cup sugar
1/4 teaspoon cinnamon
6 tablespoons cream

Preheat oven to 350°.

Grease 9x13x2-inch baking pan. In small bowl, sift 1 cup of flour, baking powder, and salt. Set aside. In medium saucepan, melt butter and chocolate over very low heat, stirring until smooth. Set aside to cool. In large bowl, combine eggs, brown sugar, and vanilla; blend. Stir in melted chocolate mixture. Gradually add flour mixture and stir until blended. Pour batter into prepared pan.

In medium bowl, combine peanut butter, sugar, cinnamon, cream, and remaining tablespoon of flour. Mixture will be stiff. Drop mixture randomly by spoonfuls on top of brownie mixture. Drag knife through peanut butter mixture to marbleize. Bake in a 350° oven for about 25 minutes. Cool before cutting.

Makes 24 brownies.

HEIDI LUTGEN
SHERIDAN, MONTANA

## SNICKERDOODLES

1 cup (2 sticks) butter
1½ cups sugar
2 eggs
2½ cups Natural White® flour
2 teaspoons cream of tartar
1 teaspoon baking soda
¼ teaspoon salt
2 teaspoons cinnamon
¼ cup sugar

Preheat oven to 350°.

Cream butter, sugar, and eggs until fluffy. Sift together flour, cream of tartar, baking soda, and salt. Add to creamed mixture. Chill dough for 30 minutes.

Mix together cinnamon and sugar. Shape dough into 1-inch balls, roll in cinnamon and sugar mixture, and place on an ungreased cookie sheet. Bake in a 350° oven for 12–15 minutes.

Makes about 4 dozen cookies.

RICHARD WEVLEY
CUT BANK, MONTANA

## OLD-FASHIONED CHOCOLATE CAKE

### Batter:
3 cups Natural White® flour
2 cups sugar
1 teaspoon salt
4 tablespoons cocoa
2 tablespoons vinegar
¾ cup vegetable oil
2 teaspoons baking soda
2 teaspoons vanilla
2 cups cold coffee

### Frosting:
½ cup (1 stick) margarine, softened
1 (8-ounce) package cream cheese, softened
1 (16-ounce) box powdered sugar
½ teaspoon vanilla

Preheat oven to 350°.

Mix all batter ingredients in a bowl. Pour batter into a 9x13-inch greased, floured pan. Bake in a 350° oven for 30–40 minutes.

Cool cake. Blend all frosting ingredients together and frost cake.

HEIDI LUTGEN
SHERIDAN, MONTANA

## STRUDEL

2¼ cups Natural White® flour
1 cup sour cream
1 cup (2 sticks) butter
3 cups pie filling

Preheat oven to 450°.

Combine flour, sour cream, and butter. Mix well to form dough. Chill dough for at least 1 hour. Roll out a quarter of the dough at a time, rolling to a ⅛-inch thick rectangle. Spread ¾ cup pie filling along one side. Roll up jellyroll fashion, very gently, as the dough tears easily. Repeat process with remaining portions of dough. Bake in a 450° oven for 18–20 minutes.

LENA CLEVELAND
BROADVIEW, MONTANA

## HAPPY HARDTACK

2 cups Bronze Chief® whole wheat flour
½ cup cornmeal
¼ cup sesame seeds, whirled in blender
½ cup Wheat Montana 7-grain cereal
½ teaspoon salt
2 teaspoons cinnamon
1 teaspoon nutmeg
⅓ cup oil
½ cup honey
¼ cup molasses
½ cup fruit juice

Preheat oven to 325°.

Combine dry ingredients. Combine wet ingredients separately, then mix both together well. Press batter into a greased 10x15-inch baking sheet with sides. The dough should fill to about ¼-inch thickness. Score with a knife before baking. Bake in a 325° oven for 45 minutes.

This is great trail food—tasty, keeps well, and provides lots of energy!

Makes 1 pan of Happy Hardtack.

PAMELA NORDHEIM
BOZEMAN, MONTANA

## PUMPKIN CAKE

### Batter:
2 cups sugar
½ cup shortening
½ teaspoon salt
2 eggs
1 cup cooked, canned pumpkin
2 cups Natural White® flour
1 teaspoon baking powder
1 teaspoon baking soda
1 teaspoon cinnamon
1 teaspoon nutmeg
⅔ cup milk
1 teaspoon vanilla
Raisins or nuts (optional)

### Powdered sugar frosting:
3–4 tablespoons soft margarine
Approximately 3 tablespoons milk
1 teaspoon vanilla or lemon flavoring
2–3 cups powdered sugar

Preheat oven to 375°.

Cream sugar, shortening, and salt until creamy. Beat in eggs; then add pumpkin. Combine flour, baking powder, baking soda, and the spices. Add milk and vanilla alternately with flour mixture to the creamed mixture. Add raisins or nuts to taste. Beat well. Pour batter into a 13x9-inch greased, floured cake pan. Bake in a 375° oven for 30 minutes. Cool cake.

Combine frosting ingredients in a bowl and mix well. Coat cooled pumpkin cake with frosting.

Serves 6.

STEPHANIE KELLOGG
CLANCY, MONTANA

## PUMPKIN COOKIES

1 cup (2 sticks) butter
1½ cups sugar
1 egg
1 cup cooked pumpkin
1 teaspoon vanilla
2½ cups Prairie Gold® whole wheat flour
1 teaspoon baking powder
1 teaspoon baking soda
½ teaspoon salt
1 teaspoon cinnamon
2 cups chocolate chips

Preheat oven to 350°.

Cream together butter and sugar. Beat in egg, cooked pumpkin, and vanilla. In a separate bowl, sift together flour, baking powder, baking soda, salt, and cinnamon. Add sifted ingredients to creamed mixture. Mix chocolate chips into batter. Place dollops of batter onto greased cookie sheets. Bake in a 350° oven for 15 minutes.

Makes 2 dozen cookies.

Pamela Nordheim
Bozeman, Montana

## RHUBARB CRISP

### Crust and Streusel:
2 cups Wheat Montana 7-grain cereal
2½ cups Natural White® flour
1¼ cups sugar
2 teaspoons baking powder
1 cup (2 sticks) margarine

### Filling:
3 cups rhubarb
1 cup sugar
1–2 tablespoons Natural White® flour

Preheat oven to 375°.

Combine 7-grain cereal, flour, sugar, and baking powder; mix well. Cut in margarine, mix until moistened. Reserve 2 cups, set aside. Press remaining mixture onto bottom of 13x9-inch baking pan. Bake in a 375° oven for 15 minutes.

Cook rhubarb and sugar on stove top until mushy. Add flour to cool mixture slightly. Pour filling over baked crust, then sprinkle reserved 7-grain cereal mixture over top, pressing lightly. Bake in a 375° oven for 30–35 minutes or until lightly golden brown.

Serves 6.

JOYCE SUTTON
LIVINGSTON, MONTANA

## GRANDMA'S APPLE CRISP

1 cup Natural White® flour
1 cup sugar (white or brown)
4 tablespoons butter
1 egg, lightly beaten
1 teaspoon vanilla
8 apples, peeled and sliced, or about 1 quart of apple slices

Preheat oven to 375°.

In a large mixing bowl, combine the flour and sugar. Using a pastry blender, cut in the butter in pea-sized portions. Add the egg and vanilla to the mixture. Mix together lightly with fork. Set this crumb mixture aside. Place sliced apples into an 8-inch, greased glass baking dish. Spread the crumbly mixture evenly over top of sliced apples. Bake in a 375° oven for 25–30 minutes or until top is browned and juice of apples is bubbling up around the edges. Serve warm with half-and-half or ice cream.

Serves 9.

JOYCE SUTTON
LIVINGSTON, MONTANA

## ALMOND CRANBERRY BISCOTTI

2¼ cups Bronze Chief® whole wheat flour
1 cup sugar
1 teaspoon baking powder
½ teaspoon baking soda
1 teaspoon cinnamon
½ teaspoon nutmeg
2 eggs *plus* 2 egg whites
1 tablespoon almond or vanilla extract
1¼ cups dried cranberries
¾ cup sliced almonds

Preheat oven to 325°.

Combine dry ingredients in a medium-sized mixing bowl. Whisk together eggs, egg whites, and extract in a separate bowl. Add to dry ingredients, mix just until moist, using an electric mixer on medium speed. Add dried cranberries and almonds; mix thoroughly.

On floured surface, divide batter in half and pat each half into a log about 14 inches long and 1½ inches wide. Place on cookie sheet. Bake in a 325° oven for 30 minutes.

Remove from oven. Reduce temperature to 300°. Cut biscotti into ½-inch slices. Place sliced side up on cookie sheet and bake an additional 20 minutes. Let cool, store in a loosely covered container.

Makes 56 biscotti.

JANET TOWNSEND
BELGRADE, MONTANA

## SUNDAY COFFEE CAKE

¾ cup sugar
1½ cups Natural White® flour
1½ teaspoons baking powder
½ teaspoon salt
1½ teaspoons cinnamon
4 tablespoons butter
½ cup milk
1 egg
Cinnamon sugar mixture to taste

Preheat oven to 350°.

Combine sugar, flour, baking powder, salt, and cinnamon; blend. Reserve ⅓ cup of mixture for later. To bulk of mixture, cut in butter. Add milk and egg to main mixture. Pour batter into greased and floured 9-inch round pan. Add additional sugar and cinnamon to taste to the reserved ⅓ cup mixture. Sprinkle on top of batter. Bake in a 350° oven for 20–25 minutes. Slice hot coffee cake and serve with butter.

Serves 6.

ELAINE HENSLEY
TOSTON, MONTANA

## NORWEGIAN PEPPER COOKIES

1 cup sugar
1 cup dark corn syrup
1 cup (2 sticks) butter
1 tablespoon vinegar
2 eggs, slightly beaten
1½ teaspoons ground black pepper
1 teaspoon ground ginger
1 teaspoon ground cloves
1 teaspoon ground cinnamon
1 teaspoon baking soda
5 cups Natural White® flour, sifted

Combine sugar, corn syrup, butter, and vinegar in a small pan and bring almost to a boil. Cool to room temperature and stir in the eggs. Sift together remaining ingredients and stir into butter mixture, blend well. Chill overnight.

Preheat oven to 350°. Divide chilled dough into several portions, and roll out each on floured board until very thin. Cut into shapes with a cookie cutter, and place on a greased baking sheet. Bake in a 350° oven for 7–8 minutes. Store cookies in a loosely covered jar in a dry place; cookies will remain crisp for a long time.

Makes about 225 (3-inch) cookies!

ELAINE HENSLEY
TOSTON, MONTANA

## AUNTIE'S SUGAR COOKIES

7 cups Natural White® flour, sifted
5 teaspoons baking powder
1 teaspoon salt
1 pound lard or shortening
2 cups sugar
4 eggs, beaten
1 cup milk

Preheat oven to 375°.

Sift together flour, baking powder, and salt. Cut lard into flour as you would for pastry, until well blended. Stir in sugar. Add eggs and milk, stir thoroughly. If dough is too dry, add more milk, a little at a time. Roll out dough on lightly floured board, cut and sprinkle with sugar. Bake in a 375° oven for 10 minutes or until done.

Makes about 21 dozen cookies.

Elaine Hensley
Toston, Montana

## COTTAGE CHEESE PIE

2 cups (skim milk) cottage cheese
1 cup sugar
¼ cup flour
1 teaspoon salt
2 eggs
¾ cup cream
Prepared pie shell
Cinnamon to taste

Preheat oven to 350°.

Stir cottage cheese until reduced into small particles. Mix with all other ingredients. Pour into pie shell and sprinkle with cinnamon. Bake in a 350° oven for 30–35 minutes.

Serves 6.

Elaine Hensley
Toston, Montana

## LEBKUCHEN (GERMAN COOKIES)

1¼ pounds granulated sugar
1 cup water
1 pint honey
4 tablespoons butter
1½ teaspoons cardamon
5 beaten egg yolks or 2 whole eggs
6–8 cups Natural White® flour
1 teaspoon baking soda
Walnuts or almonds (optional)

Combine sugar and water and boil until it spins a thread. Remove from heat. Add honey, stirring constantly. Mix in butter, cardamon, eggs, flour, and baking soda. Be careful not to add too much flour. Dough must be pliable. If desired, add walnuts or almonds. Let rise overnight, covered.

Preheat oven to 350°. Roll out dough, cut out cookies into desired shapes and sizes, and place on a cookie sheet. Bake in 350° oven for 10–12 minutes, or until lightly browned.

Store baked cookies in airtight containers for weeks to ripen before they are good; they stay soft.

This is an old, old German recipe!

ELAINE HENSLEY
TOSTON, MONTANA

## PIONEER CAKE

3 cups Natural White® flour
2 cups sugar
2 teaspoons soda
1/2 teaspoon salt
6 tablespoons cocoa
3/4 cup vegetable oil
3 tablespoons vinegar
3 tablespoons vanilla
3 cups cold water
Sugar for sprinkled topping (optional)

Preheat oven to 375°.

Sift first five ingredients into ungreased 10x14-inch pan. Mix and make 3 "wells." Into one well add oil, into other well add vinegar, and into last well add vanilla. Pour cold water over all. With fork, mix together. Sprinkle with sugar if desired. Bake in a 375° oven for 35–40 minutes. If you didn't sprinkle with sugar before baking, you can frost cooled cake with your favorite frosting.

DEENA GOBBS
TOWNSEND, MONTANA

## WHOLE WHEAT PIE CRUST

2½ cups Bronze Chief® or Prairie Gold® whole wheat flour
1 cup (2 sticks) chilled margarine or butter
1 teaspoon salt
1 tablespoon vinegar
1 egg, beaten
6–8 tablespoons ice water

Preheat oven to 425°.

Mix the flour, butter, and salt with a mixer or by hand until the mixture is the size of large peas. With a fork, combine the vinegar, egg, and ice water. Blend together all ingredients. Divide dough into 3 equal portions; put 2 in the refrigerator. Roll out 1 into an 8-inch crust and put in pie pan. Do the same with remaining portions, keeping one chilled while the other is being rolled. Bake in a 425° oven for 8 minutes.

Makes 3 single crusts.

BETTY HORNE
TOWNSEND, MONTANA

## QUICK CARAMEL PEACH PUDDING

1 cup Natural White® flour
½ cup sugar
½ teaspoon salt
2 teaspoons baking powder
1 cup milk
¼ teaspoon almond extract
2 cups sliced fresh peaches

**Caramel Topping:**
½ cup white sugar
¼ teaspoon nutmeg
2 cups brown sugar
1¼ cups water
1 tablespoon butter

Preheat oven to 400°.

Mix flour, sugar, salt, and baking powder. Stir in milk and almond extract. Fold in peaches. Spread mixture in an 8- or 9-inch square pan.

Combine topping ingredients in saucepan and heat until dissolved. Pour heated caramel topping over batter and bake in a 400° oven for 30 minutes.

Serve warm with cream or ice cream.

Serves 4–6.

LENA CLEVELAND
BROADVIEW, MONTANA

## PRAIRIE GOLD® RANGER COOKIES

1 cup white sugar
1 cup brown sugar
3 eggs, beaten
1 cup shortening
1 teaspoon vanilla
$\frac{1}{2}$ teaspoon baking powder
1 teaspoon baking soda
$\frac{1}{2}$ teaspoon salt
2 cups Prairie Gold® whole wheat flour
2 cups cornflakes
1 cup coconut
2 cups Wheat Montana 7-grain cereal
2 cups chocolate chips
$\frac{1}{2}$ cup sunflower seed meats

Preheat oven to 350°.

Cream together first 5 ingredients until fluffy. Sift together next 4 ingredients and stir into creamed mixture. Add remaining ingredients to mixture. Use hands to blend in; mixture will be stiff.

Form dough into balls about $1\frac{1}{2}$ inches in diameter. Bake on ungreased cookie sheets in a 350° oven for 10–15 minutes.

# Soups & Salads

*Chicken Pot Pie Soup*
*(see page 140)*

WHEAT MONTANA
THREE FORKS, MONTANA

## WHEAT SALAD

1 (8-ounce) package cream cheese, softened
1 (6-ounce) package vanilla instant pudding
3 tablespoons lemon juice
1 (15-ounce) can crushed pineapple with juice
8 ounces Cool Whip
1½ cups Cooked Wheat (see page 144)

Combine cream cheese with pudding mix and mix well. Stir in lemon juice and pineapple with juice. Fold in Cool Whip. Stir well, gently stirring in Cooked Wheat in order to keep kernels whole. Chill.

Can be prepared ahead of time. Keeps several days in refrigerator.

Serves 4–6.

WHEAT MONTANA
THREE FORKS, MONTANA

## BERRY SUPREME SALAD

4 hard-boiled eggs, chopped
4 sweet pickles, chopped
½ cup chopped onion
½ cup diced Cheddar cheese
1 cup diced celery
1 cup mayonnaise
4 cups Cooked Wheat (see page 144)
2 tablespoons chopped pimento
1 cup diced lunch meat or cooked ham
Salt to taste

Combine all ingredients, mix well, and refrigerate at least 4 hours before serving.

Serves 6–8.

WHEAT MONTANA
THREE FORKS, MONTANA

## WHEAT BERRY & MUSHROOM SALAD

**Marinade:**
1/4 cup oil
1/4 cup white vinegar
1 clove garlic, minced
1 tablespoon Dijon mustard
1 1/2 teaspoons oregano
1/4 teaspoon salt
1/8 teaspoon pepper

**Salad:**
3 cups Cooked Wheat (see page 144)
1/4 pound fresh or canned mushrooms
1 small can black olives
1/4 cup chopped green pepper

Mix together marinade ingredients and let sit overnight.

Combine salad ingredients and add to marinade.

WHEAT MONTANA
THREE FORKS, MONTANA

## SEAFOOD BERRY SALAD

1 large, ripe pineapple
3 cups Cooked Wheat (see page 144)
½ cup snipped parsley
1 cup ripe olives, chopped
4 green onions, sliced
⅓ cup mayonnaise
3 tablespoons French dressing
½–1 teaspoon curry powder
¾ cup diced celery
1 cup seafood (crab, shrimp, tuna) or 1 cup turkey or chicken
Salt and pepper to taste

Split pineapple in half lengthwise. Hollow out shells. Cube pineapple meat;
reserve. Chill shells and cubed pineapple. Combine all other ingredients except
seafood, salt and pepper. Chill for several hours or overnight. Just before serving,
add seafood and 1 cup of cubed pineapple. Season to taste; pile salad into hollow
shells and serve.

Serves 8.

FRANCES FOLKVORD, WHEAT MONTANA
THREE FORKS, MONTANA

## GRANDMA FRAN'S "COOL SALAD"

    1 cup cooked string beans (measure after cooking)
    1 cup diced cooked carrots (measure after cooking)
    1 cup Cooked Wheat (see page 144)
    1 cup diced Cheddar cheese
    1 cup diced cooked ham
    4 hard-boiled eggs, diced
    2 cups Romaine lettuce, chopped
    2 tomatoes, diced

Combine all ingredients and mix with French dressing to taste. Serve on lettuce leaf.

Serves 8.

WHEAT MONTANA
THREE FORKS, MONTANA

## POTATO & WHEAT SOUP

2 large potatoes, diced
1 tablespoon onion, minced
1 cup milk
1 cup water
1 (10¾-ounce) can cream of potato soup
2 tablespoons butter
½ cup Cooked Wheat (see page 144)

Cook potatoes and onion in milk and water. Add potato soup, butter, and wheat. Simmer 30 minutes.

Serves 4.

WHEAT MONTANA
THREE FORKS, MONTANA

## HEADWATERS WHEAT CHILI

2 cups wheat berries
7 cups water
1½–2 pounds browned hamburger (drained well)
2 (16-ounce) cans stewed tomatoes with juice
1 (16-ounce) can tomato sauce
2 tablespoons chili powder
¼ teaspoon cumin
Chopped green peppers, onion, celery to taste (optional)

The night before you wish to eat the chili, combine wheat berries and water in a Crockpot. Cook overnight on low.

Add remaining ingredients and continue simmering for 3–4 hours.

Makes 10 healthy servings.

FRANCES FOLKVORD, WHEAT MONTANA
THREE FORKS, MONTANA

## WHEAT CHILI

>2 cups Prairie Gold® wheat berries
>3 cups water
>1 pound extra lean ground beef
>1 large onion, chopped
>1 green pepper, chopped
>1 teaspoon salt
>1 teaspoon chili powder
>⅛ teaspoon cayenne pepper
>⅛ teaspoon black pepper
>1 (8-ounce) can tomato paste
>1 (8-ounce) can whole tomatoes
>4 cups beef broth

Cook wheat berries in water for 1 hour until tender; add more water if necessary. Drain and rinse with cool water. In a skillet over medium heat, brown beef and onion; drain in a colander and rinse with hot water. Stir in remaining ingredients. Simmer, uncovered, 1 hour or until desired consistency is reached. Stir occasionally.

Serves 6.

MARGARET CASTER
WORDEM, MONTANA

---

## POT PIE SQUARES

2 eggs
2 cups Natural White® flour
2–3 tablespoons milk or cream

Break the eggs into the flour and work together. Add the milk to make a soft dough. Roll out the dough as thin as possible and cut into 1x2-inch rectangles with a knife or pastry wheel. Cook in Chicken Pot Pie Soup below.

## CHICKEN POT PIE SOUP

1 (3½–4-pound) chicken
4 medium-sized potatoes, peeled and cut into chunks
1 onion, diced
Salt and pepper to taste
Pot Pie Squares (see above)

Cook the chicken in 2 quarts water until it is partly tender. Then add the potatoes and onion. Cook until vegetables and chicken are completely tender. Retain broth with vegetables. Remove meat from bones and set aside. Bring broth to a boil. Drop Pot Pie Squares into boiling broth and cook 20 minutes or until tender. Return chicken to the broth and serve steaming hot.

Serves 6–8.

MARGARET CASTER
WORDEM, MONTANA

## RIVVELS

¾ cup Natural White® flour
1 egg

Put flour in bowl. Break egg into flour and mix with a fork until dry and crumbly. Crumble this mixture into the soup (below) and stir occasionally so that rivvels separate.

## CHICKEN CORN SOUP

1 (3–4 pound) fryer chicken, cut up
Salt to taste
2 quarts corn (frozen, fresh, or canned)
Rivvels (optional, see above)
3–4 hard-boiled eggs, diced
Pepper to taste

In large kettle, cover chicken pieces with water. Salt to taste, cook until tender. Cut meat off bones and dice into bite-sized pieces. Return chicken to broth, add corn and return to a boil. Stir in rivvels and eggs and cook until rivvels are cooked through and float on the soup's surface. Add pepper and serve.

Makes 8–10 servings.

JOYCE SUTTON
LIVINGSTON, MONTANA

## WHEAT SALAD

1½ cups uncooked Prairie Gold® or Bronze Chief®
wheat berries
1 (8-ounce) package cream cheese
1 (15–16-ounce) can crushed pineapple, in juice
1 (5-ounce) package instant pudding, pistachio flavored
3 tablespoons lemon juice
1 (12-ounce) container whipped topping

Cook wheat berries in 6 cups water. Bring to boil, turn to low or simmer; cook
for 1 hour. Check berries for tenderness. If not tender, continue cooking on low
until easy to chew. At any time during cooking, you may add water so the berries
don't cook dry. When the berries are tender and easy to chew, remove from heat.
Drain in colander, wash with cold water, then drain until all moisture is gone.
You can cook the berries the day before needed, then refrigerate in container
with tight-fitting lid.

In a large mixing bowl, combine cream cheese with crushed pineapple and juice,
dry pudding, and lemon juice. Add cooled Cooked Wheat, mix well, add
whipped topping last. Transfer mixture to a smaller container with tight fitting
lid. Can be refrigerated up to a week.

This is either a dessert or a side salad. It is great for potlucks.

Makes 12-15 servings.

# Miscellaneous

*Cheese Straws*
*(see page 162)*

WHEAT MONTANA
THREE FORKS, MONTANA

## COOKED WHEAT

1 cup cleaned, rinsed, raw Bronze Chief® or Prairie Gold®
wheat kernels
3 cups water

Combine ingredients, cover, and let sit overnight or for 12 hours. Do not drain. Place soaked wheat over heat and boil about 5 minutes, then simmer until tender, about 30 minutes.

Or, combine wheat kernels and water and place in a 225° oven for 6 hours. Keep covered with water. Rinse until water is clear. Drain.

Store in freezer in plastic zip-locked bags. Frozen wheat will last more than 1 year. Cook large batches and freeze for later use.

1 cup uncooked wheat=2 cups Cooked Wheat.

WHEAT MONTANA
THREE FORKS, MONTANA

## RANCH WHEAT CASSEROLE

1 cup Cooked Wheat (see page 144)
1/2 cup tomato paste or tomato juice
2 tablespoons molasses
2/3 cup water
2 tablespoons brown sugar
1/2 teaspoon dry mustard
1/2 teaspoon salt
1–2 tablespoons onion, minced
4 slices bacon

Combine first 8 ingredients and top with bacon if desired. Cover and bake 45 minutes in a preheated 325° oven. Uncover and bake an additional 40 minutes.

Serves 4.

WHEAT MONTANA
THREE FORKS, MONTANA

## INSTANT CEREAL

3 cups water
1⅓ cups Cooked Wheat (see page 144)
½ teaspoon salt
1 (32-ounce) insulated jug

Combine water, Cooked Wheat, and salt in a saucepan. Bring to a boil; simmer for 5 minutes. Pour into an insulated bottle or picnic jug. Close. Let stand overnight. In the morning, the wheat is tender, ready to use, and even hot! Add milk and sweetener for a tasty "instant" breakfast.

Serves 5.

Wheat Montana
Three Forks, Montana

## SWEET & SOUR WHEAT CASSEROLE

1 pound hamburger
1 teaspoon Johnny's seasoning salt
1 cup onion, chopped
3 cups Cooked Wheat (see page 144)
1 cup diced celery
1 (8-ounce) can tomato sauce
½ cup water
⅓ cup brown sugar
¼ cup catsup
1 tablespoon vinegar
1 tablespoon prepared mustard

Preheat oven to 350°.

In a frying pan, brown hamburger with seasoning salt and onion. Add other ingredients. Combine and put into a 9x12-inch casserole dish. Bake in a 350° oven for 1½ hours.

WHEAT MONTANA
THREE FORKS, MONTANA

## CREAM OF WHEAT CASSEROLE

2 cups Cooked Wheat (see page 144)
1 tablespoon dried minced onion
2 tablespoons chopped pimento
1 (10¾-ounce) can cream of chicken soup
1 cup sour cream
5 ounces cubed Velveeta cheese
Dash pepper
1 (2.8-ounce) can French-fried onion rings

Combine and mix first 7 ingredients and bake in a preheated 350° oven until bubbly and brown. Top with French-fried onion rings and bake 10 minutes longer.

Serves 4–6.

WHEAT MONTANA
THREE FORKS, MONTANA

## VERY VEGGIE

2 large carrots, chopped
2 cups Cooked Wheat (see page 144)
1 head cauliflower, separated into flowerets
2 (10¾-ounce) cans cream of mushroom soup
2 teaspoons celery salt
1 teaspoon pepper

Bring to a boil the carrots and Cooked Wheat. Cook 20 minutes. Add cauliflower and cook 15 minutes longer. Combine remaining ingredients and serve.

Serves 6.

WHEAT MONTANA
THREE FORKS, MONTANA

## WHEAT DUMPLINGS

½ cup Cooked Wheat (see page 144)
4 tablespoons chicken or turkey broth
1 egg
½ teaspoon poultry seasoning (or sage, parsley, thyme, salt, etc.)
1 cup Natural White® flour
2 teaspoons baking powder

Combine Cooked Wheat, broth, and egg in a blender and beat well. In a separate bowl, combine seasoning, flour, and baking powder. Mix and drop into broth mixture. Cook uncovered over medium heat in skillet or large saucepan for 10 minutes, then cook covered for 10 minutes.

Serves 4.

WHEAT MONTANA
THREE FORKS, MONTANA

## WHEAT PILAF

4 cups Cooked Wheat (see page 144)
3 tablespoons butter or margarine
1 (1½-ounce) package dry onion soup mix
2 cups water
1 (4-ounce) can sliced, drained mushrooms

Sauté Cooked Wheat in butter. Add remaining ingredients. Place in greased 1½-quart casserole dish. Bake in a preheated 350° oven for 1½ hours. Check, and add more water if needed.

WHEAT MONTANA
THREE FORKS, MONTANA

## WHEAT PUDDING

2 cups Cooked Wheat (see page 144)
3 eggs, slightly beaten
1 tablespoon grated orange rind
2½ cups milk
⅓ cup honey
⅛ teaspoon salt
Fruit for topping

Blend all ingredients except fruit. Pour into a buttered 2-quart baking dish. Bake in a preheated 350° oven for 1 hour or until set. Serve warm, topped with berries, applesauce, or sliced fruit.

Serves 4.

WHEAT MONTANA
THREE FORKS, MONTANA

## WHEAT SNACKS

Place drained Cooked Wheat (see page 144) on paper towel. Salt lightly. Place in shallow pan in oven for 10 minutes. Preheat oil to 375° in small deep pan (not an electric skillet). Carefully place small amount of Cooked Wheat in oil (moisture causes oil to bubble). When kernels rise to the surface, remove with slotted spoon and place on paper towels to absorb excess oil. Season with salt, seasoned salt, or flavored salts such as ham, smoke, or hickory. Grated Parmesan cheese may also be used for flavor. To make crispier, place fried wheat in shallow pan and bake in a 350° oven 8–10 minutes. Store in a cool, dry place.

WHEAT MONTANA
THREE FORKS, MONTANA

## WHEAT MONTANA 7-GRAIN CEREAL
## WITH APPLES AND NUTS

> 2 cups Wheat Montana 7-grain cereal
> 2 medium apples, diced
> 1/2 cup walnuts or almonds
> 1 teaspoon cinnamon
> 2 tablespoons maple syrup or honey

Cook cereal per package directions. Stir in remaining ingredients.

Serves 4.

WHEAT MONTANA
THREE FORKS, MONTANA

## WHEAT RELISH

6 cups Cooked Wheat (see page 144)
1½ cups vinegar
1 cup oil
2 tablespoons sugar
1½ teaspoons garlic salt
1 clove garlic, minced
1 cup celery, diced
1 cup onion, chopped
¾ teaspoon basil
¼ teaspoon tarragon
¼ cup dried parsley
½ teaspoon dill weed
¼ teaspoon pepper
2 teaspoons salt
2–3 tablespoons dry, mixed dehydrated vegetables

Mix all ingredients thoroughly. Refrigerate or process in hot water bath as for pickles. Excellent on roasts, fish, baked potatoes, salads, and sandwiches. Will store refrigerated for 2–3 weeks.

Makes 6–7 cups.

WHEAT MONTANA
THREE FORKS, MONTANA

## SOUTH OF THE BORDER

1 pound hamburger
½ cup chopped onion
½ cup chopped green pepper
3 cups Cooked Wheat (see page 144)
1 tablespoon chili powder
1 (15-ounce) can creamed corn, drained
1 can whole kernel corn, drained
1 (15-ounce) can tomato sauce
1 teaspoon garlic salt
1 cup shredded Cheddar cheese

Sauté hamburger, onion and green pepper in a skillet over medium heat until browned and cooked through. Drain grease and add Cooked Wheat, chili powder, corn, tomato sauce, and salt.

Bake in a preheated 350° oven for 30 minutes. Put shredded Cheddar cheese on top. Bake 15 minutes more. Makes great tacos.

Serves 4–6.

WHEAT MONTANA
THREE FORKS, MONTANA

## WHEAT & MEAT CASSEROLE

1½ pounds hamburger
¼ cup chopped onion
1 cup chopped celery
1 cup Cooked Wheat (see page 144)
1 (8-ounce) can tomato sauce
½ cup catsup
¼ teaspoon pepper
1 teaspoon *plus* ¼ teaspoon salt
2 cups mashed potatoes
1 beaten egg
1 tablespoon flour

Brown meat, onion, and celery in a skillet over medium heat. Add Cooked Wheat, tomato sauce, catsup, pepper, and 1 teaspoon salt to the skillet. Combine. Put browned meat mixture into a 9x12-inch baking dish.

Combine mashed potatoes, egg, ¼ teaspoon salt, and flour, and add to the top of casserole. Bake in a preheated 350° oven for 1 hour.

Serves 4.

WHEAT MONTANA
THREE FORKS, MONTANA

## WHOLE WHEAT ZUCCHINI QUICHE

  1 pie crust, baked
  1¼ cups Cooked Wheat (see page 144)
  2 eggs
  1 egg yolk
  1 cup peeled, cooked, mashed zucchini
  1½ cups milk
  ½ teaspoon salt
  8 ounces Cheddar cheese, shredded

Preheat oven to 325°.

Line bottom of crust with Cooked Wheat. Combine remaining ingredients and pour on top of wheat. Bake in a 325° oven for 40 minutes.

Serves 6.

WHEAT MONTANA
THREE FORKS, MONTANA

## WHOLE WHEAT CASSEROLE

1 pound hamburger, bacon, or sausage
1/2 cup chopped onion
2 cups Cooked Wheat (see page 144)
1/4 teaspoon pepper
1 (8-ounce) can tomato sauce
1/2 cup catsup
1/2 teaspoon salt
1/2 teaspoon garlic salt
2 (8-ounce) cans tomatoes, diced
Dash liquid smoke or 1 tablespoon Worcestershire sauce

Preheat oven to 350°

Brown meat and onion in a frying pan. Drain and add other ingredients. Place in greased 9x12-inch casserole dish. Bake in a 350° oven for 1 1/2–2 hours.

Serves 4–6.

Diana Hjertberg
Big Timber, Montana

## AXELROD & REBA'S PEANUT BUTTER DOG BISCUITS

3 cups Prairie Gold® whole wheat flour
½ cup rolled oats
2 teaspoons baking powder
1½ cups milk
1¼ cups peanut butter
1 tablespoon molasses

Preheat oven to 350°.

Combine the flour, oats, and baking powder in a large bowl. Using a food processor or blender, mix the milk, peanut butter, and molasses until smooth. Add to the dry ingredients. Using your hands, knead the ingredients together. Dough will be quite stiff.

Roll out the dough to ¼-inch thickness and cut with cookie cutters. Bake in a 350° oven for 20–25 minutes or until lightly browned. Turn off the heat and leave the biscuits in the oven until cool. Store in airtight container.

MARILYN STEINGRUBER
MANHATTAN, MONTANA

## CHEDDAR PARMESAN POTATOES

4 tablespoons butter or margarine
1/4 cup Natural White® flour
2 cups milk
1/2 teaspoon salt
1 cup shredded Cheddar cheese
1/2 cup grated Parmesan cheese
5 cups sliced potatoes
1/4 cup buttered bread crumbs

Preheat oven to 350°.

In a saucepan, melt butter over low heat. Stir in flour until smooth. Gradually add milk; cook and stir over medium heat until mixture thickens. Remove from heat. Add the salt, Cheddar cheese, and Parmesan cheese. Stir mixture until cheeses melt. Add potatoes; stir gently to mix. Place in greased 2-quart baking dish. Sprinkle bread crumbs on top. Bake uncovered in a 350° oven for 30–35 minutes.

Makes 6–8 servings.

ELAINE HENSLEY
TOSTON, MONTANA

## CHEESE STRAWS

¾ pound sharp Cheddar cheese, grated
¾ cup (1½ sticks) margarine, softened
2½ cups Natural White® flour
¾ teaspoon salt
⅜ teaspoon cayenne pepper

Preheat oven to 375°.

Blend cheese and margarine. Gradually add dry ingredients. Flatten dough to ¼-inch thick; refrigerate 1 hour. Then cut into thin strips and twist.

Bake in a 375° oven for 10 minutes. Don't brown. Store in an airtight container.

ELAINE HENSLEY
TOSTON, MONTANA

## TOAD IN THE HOLE

1 slice Wheat Montana Bread (Montana Toast works well)
Margarine
1 egg
Seasonings to taste

Tear a hole about the size of a silver dollar out of the middle of a slice of Wheat Montana bread. Using a small amount of margarine in a frying pan, toast one side of bread and turn over. Break an egg into the hole, season to taste and turn over to finish cooking egg, using margarine. May need to flip again to reach desired egg firmness.

ELAINE HENSLEY
TOSTON, MONTANA

## SOURDOUGH STARTER

1 cup warm water
1¼ cups *plus* 1 teaspoon Natural White® flour
1 teaspoon sugar
1 teaspoon salt
1 medium potato, raw and grated

Mix all ingredients and place in glass jar or crock. Cover with cheesecloth, so wild yeast in the air can settle on it for 24 hours. Then place lid (not sealed) on jar and allow to ferment for 2 weeks. Keep in covered jar at temperature less than 70°.

Follow your favorite recipe for baking bread or pancakes using sourdough starter.

Always add same amount of flour and water to jar as amount you take out for recipe.

Marilyn Steingruber
Manhattan, Montana

## WAFFLE SAUCE

2¾ *plus* ¼ cups milk
⅔ cup sugar
3 tablespoons Natural White® flour
1 tablespoon butter
1 teaspoon vanilla

Heat 2¾ cups milk; mix in sugar and flour. Stir in the remaining ¼ cup milk. Boil mixture until thickened. Add butter and vanilla. Pour over hot waffles.

## Wheat

**Bronze Chief® Hard Red Spring Wheat Berries (50 lb. bags, 45 lb. pails)** — High Protein Hard Red Spring Wheat. This is the wheat that founded our business. Naturally air dried, plump kernels, low moisture, and excellent baking quality. For superior whole-wheat baking. Chemical free.

**Prairie Gold® Hard White Spring Wheat Berries (50 lb. bags, 45 lb. pails)** — This is the wheat that started a revolution in bread baking. Since we first started selling Prairie Gold® in 1988, thousands of home, specialty, and commercial bakers have discovered the unique characteristics of this grain. Its naturally golden color makes 100% whole-wheat baked foods that are lighter and sweeter. It is excellent in all bread recipes, cookies, pie crusts, and more. Naturally air-dried to low moisture. Chemical free.

**Certified Organic Hard Red Spring Wheat Berries (50 lb. bags)** — *Certified organic* means this grain has been grown on land that has had no chemicals or fertilizer applied for 4 years. The protein is a bit lower than our chemical-free grains, and the baking consistency may vary. However, for those customers who demand certified organic status, this grain is perfect in every way.

**Soft White Wheat Berries (50 lb. bags)** — Soft white wheat is generally used for cookies, crackers, and other baked goods that do not require a high gluten content. This grain is low in protein. It is not intended for making great bread, but it will perform well in most other baking applications. Chemical free.

**Bronze Chief® Cracked Wheat (25 lb. bags)** — Once cracked, wheat becomes much more versatile for use as an additional bread ingredient or soaked and cooked in wheat recipes. If cooked and dried, it becomes bulgar. There are lots of ways that cracked wheat can add versatility to your recipes. Chemical free.

**Wheat Bran (50 lb. bags)** — When unbleached white flour is made, the bran and germ of the wheat kernel are removed, leaving only the white starchy endosperm which is finally ground for white flour. The bran becomes a by-product in the process. The bran layer contains the highest concentration of nutrition and vitamins, plus soluble fiber. Our bran is excellent and can be used alone or as an addition to a variety of baking applications. Chemical free.

# Flour

**Natural White® Unbleached White Flour (50,10, and 5 lb. bags)** – This is the highest protein unbleached white flour available. You will find it to be the best white baking flour you've ever used. Carefully milled to exact specifications and enriched with B-vitamins, niacin, and iron. We also add a small amount of malted barley to this blend for further enhancement of final product consistency. Whether you are baking at sea level or high altitude, this flour will perform to your highest expectations. Chemical free.

**Bronze Chief® Whole Wheat Flour (50, 10, and 5 lb. bags)** – This is our own Bronze Chief® wheat ground finely, with your baking interests in mind. By milling this flour ourselves, we control the speed of the grind. We go slow and make sure that heat gain during the process doesn't destroy the protein, vitamins or baking quality. We believe that this is the highest quality, traditional whole-wheat flour available in the world. 100% whole-wheat flour. Chemical free.

**Prairie Gold® Whole Wheat Flour (50, 10, and 5 lb. bags)** – Literally thousands of bakers have told us that they wouldn't use anything else. This flour, ground from our Prairie Gold® Wheat, will delight you, your family, and your customers in every way. It's so versatile it can be substituted for white flour in most recipes, giving your baked goods unique flavor, plus all of the benefits of whole wheat. 100% whole-wheat flour. Chemical free.

## Specialty Grains

**7-grain Specialty Mix (25 lb. bags, 45 lb. pails)** – This unique blend of seven whole grains covers all of the bases: protein, nutrition, vitamins, fiber, and flavor. Made from a hearty combination of hard wheat, soft wheat, triticale, rye, oats, pearled barley, and spelt. You can use it whole, soak it, crack it, or grind a flour suitable for stand-alone baking. Chemical free.

**Rolled 7-grain Specialty Mix (50 lb. and 3 lb. bags)** – This mix is ready to eat. It's so easily digestible, you can eat it by the handfuls right out of the bag. You'll enjoy these rolled flakes either as a bread ingredient, topping, or cereal (cold or hot). We've sold thousands of bags of the mix to customers across the nation who continue to find new uses for one of our most versatile and popular products.

**Cracked 7-Grain Specialty Mix (50 lb. bags)** – The same combination of seven grains, except we've cracked it for you. This mix can be used as is, or further processed into flour. Chemical free.

**Cracked 9-Grain Specialty Mix (50 lb. bags)** – Similar to our 7-Grain mix with the addition of durum wheat and flax. If seven grains isn't enough, this mix should do it. Most folks report a nuttier flavor with this unique blend. Bakeries usually like this one better. Chemical free.

**Spelt Berries (50 lb. bags)** – The origin of spelt has been traced all the way back to the first cultivations of grain, around 500 B.C. Modern day farmers grew spelt mostly as livestock feed, because once harvested, it is protected by a fibrous hull that is very difficult to remove. Today, however, it has found its niche in the human food industry as a substitute for wheat or other gluten-containing grains. Many who have allergic reactions to regular grains are able to consume spelt and still enjoy a diet with grain protein and soluble fiber. This grain is a little tougher to bake with, but with some trial and error, most bakers manage predictable results in a full spectrum of gluten-free baked products. Chemical free.

**Rye Berries (50 lb. bags)** – Rye bread is most popular in Europe, where it is the predominant bread consumed. However, rye bread is gaining acceptance in America at a rapid pace. Rye, when used as a baking ingredient, imparts a flavor that is all its own. We clean and package only the best berries, so you can expect success no matter how you use them. Chemical free.

**Kamut Berries (50 lb. bags)** – This is an ancient relative to wheat. You'll enjoy these thick, long-grained, bronze kernels. Kamut is high in protein and easily digestible. It is currently being used in cereal, crackers and bread products nationwide. Licensing agreements keep this product in limited supply. Kamut is certified organic.

**Pinto Beans (25 lb. bags, 45 lb. pails)** – Our beans are cleaned and sized to perfection and low in moisture content, making them the best available at any price. Chemical free.

**Navy Beans (small white) (25 lb. bags, 45 lb. pails)** – Navy beans are versatile, high in protein, and healthy. There is a tremendous amount of food value in a small quantity of beans. Plus, the uses are endless. Chemical free.

**11-Bean/Pea Mix (25 lb. bags, 45 lb. pails)** – This is one way to have 'em all. Beans and peas that is. This mixture includes lentils, whole green peas, split yellow and green peas, black-eyed peas, baby lima beans, pintos, and small white and small red navy, black turtle, and pink beans. You can take it from here for great soups, salads, or whatever. Chemical free.

**Pearled Barley (25 lb. bags, 45 lb. pails)** – Pearling the barley actually removes the hull and germ, plus the outer layer of the kernel itself. Pearled barley is more appropriate for soups, soaking, or barley flour. A small amount of barley flour, added to regular wheat flour, can actually improve baking consistency, while adding beta glucans that enhance health. Chemical free.

**Hulled Barley (25 lb. bags, 45 lb. pails)** – Barley is a grain that has a fibrous hull tightly attached to the kernel. By hulling the barley, we remove the indigestible part and leave the rest. This grain is ready for cracking or grinding

**Yellow Corn (25 lb. bags, 45 lb. pails)** – Whole, cleaned, #1 yellow corn. Makes excellent meal or corn flour. Add to recipes or use as a stand-alone bakery ingredient. Chemical free.

**Alfalfa (25 lb. bags, 45 lb. pails)** – Alfalfa, when grown as a crop, produces a heavily leafed forage for livestock. However, everyone can attest to the nutritional qualities and flavor profile of the sprouted seeds. Ours are high in germination and cleaned to perfection. These small seeds are dynamite. Chemical free.

**Lentils (25 lb. bags, 45 lb. pails)** – These round, flattened seeds are awesome when sprouted, soaked, or used in soup recipes. There are many uses for lentils, some of which are still being discovered today. These are rated #1 green lentils. Chemical free.

**Sunflower Seeds (50 lb. bags)** – Ours are the shelled, raw seeds. They are certainly versatile as a recipe ingredient, topping, or roasted. Chemical free.

**Buckwheat Groats (50 lb. bags)** – Buckwheat, when harvested, is protected by a small triangular hull inedible by humans. By groating the seed, the hard hull is removed, leaving the edible, flavorful groat. We package only the best part, so that you get what you expect. Chemical free.

**Millet (50 lb. bags)** – Widely cultivated in the old world, millet is seeing a resurgence in our modern-day food supply. This grass plant produces a white seed that is high in protein and certain trace minerals. Ours is hulled and ready to use. Chemical free.

**Oat Groats (50 lb. bags)** – Oats contain a heavy hull after harvest. By groating the kernel, we remove the hull, leaving just the most palatable part. Our oat groats are soft, chewy, and packed full of flavor. Rest assured, oats are among the most nutritionally beneficial of the grains available today. Our groats are produced from select, sound, clean, hulled, heavy test weight, white oats. Chemical free.

**Rolled Oats (50 lb. bags)** – Rolled oats have been and continue to be a mainstay in the modern diet. High in healthy fiber and nutrition, and made more versatile and digestible by the rolling process. Our rolled oats are table grade resulting from the flaking of cleaned, steamed groats. You can't beat our rolled oats for quality. Chemical free.

**Long Grain Brown Rice (25 lb. bags, 45 lb. pails)** – Brown rice is an unrefined, highly nutritious food staple that provides the basic food needs for more of the world's population than any other grain. This is premium quality with less than 4% broken kernels. Chemical free.

**Long Grain White Rice (25 lb. bags, 45 lb. pails)** – The same high quality as our brown counterpart, but with most of the dark-colored hull removed. Enriched with B-vitamins, niacin, and iron. Chemical free.

## Specialty Products

**Pure Montana Honey (12 lb. pail, 60 lb. pail)** – Our honey is produced at the base of the Crazy Mountains of central Montana. The bee's primary source of nectar is sweet clover and alfalfa blooms. Some honey is dark and strong flavored. This honey is light amber and almost white in color, with a mild, sweet flavor. It has been warmed, strained, and then packed in a hermetically sealed plastic bucket. Our honey will always maintain its flavor and usability.

**Dried Milk Powder, Nonfat (55 lb. bag)** – Dried milk powder results from the removal of the fat and water from fresh sweet milk. This powder still contains all the lactose, milk proteins, and milk minerals in the same relative proportions as they occur in fresh milk.

**Plastic Pails** – Our 6-gallon pails, with high-strength lids, provide a safe haven for all grains and food products. The lids are "knock-on" type with a hermetic seal that guarantees the integrity of its contents. An all-metal handle is included for easy maneuvering and stacking.

**Pail Opener** – Once the lid is secured, it is hard to get off. Many people use a screwdriver or have rigged up their own opening tools. We have gone to our pail supplier for a factory-made pail opener that fits under the lip and easily removes the lid. It's much easier than any homemade tool we've seen and takes the hassle out of opening pails.

**NOTE:** All products packaged in 45 lb. plastic pails contain an oxygen-absorbing packet.

The term *berries* refers to the whole kernel form of the grain.

All grains store best in cool, dry conditions. Bugs will infest grain and flour if heat and humid conditions exist. Keep your products cool and dry.

Our facilities are USDA, FDA, US Army and State of Montana inspected and certified as a food quality processing plant. We are inspected and certified to process and package Certified Organic products.

We have many questions regarding the "chemical free" status of our wheat and wheat products. Every year, and usually after harvest, we submit representative wheat samples to the State Chemical Testing Laboratory. The lab puts each sample through a set of 36 different chemical residue tests. The grain is tested for chemical residue to the parts per million in each of the 36 test categories. If the test results confirm that the grain has no detectable amount of chemicals, and if it meets our milling and baking quality criteria, then it can be labeled as "chemi-cal free" and processed further. We hope that this extra effort in quality assurance on our part gives our customers the knowledge that they are purchasing safe, high-quality grain and flour products.

**Bronze Chief®, Prairie Gold® and Natural White®** are registered trade names of Wheat Montana Farms, Inc.

For more information about all these products, contact Wheat Montana at 1-800-535-2798 or visit their website at www.wheatmontana.com.

# -Index-

# Z